T0174794

The Mediterranean Region

First published in 1984, *The Mediterranean Region* considers the broad economic and political problems facing the region from a variety of perspectives.

The book features the work of international experts on the Mediterranean region. It discusses the changing legal environment and covers the Law of the Sea as applied to the Mediterranean, and the position of the Arab countries in the region. Chapters are also devoted to exploring the different policies of Russia, the United States, and Europe, towards the Mediterranean.

The Mediterranean Region is a detailed contribution to research and understanding of the area.

The Mediterranean Region

Economic Interdependence and the Future of Society

Edited by Giacomo Luciani

Routledge
Taylor & Francis Group

First published in 1984
by Croom Helm Ltd.

This edition first published in 2021 by Routledge
2 Park Square, Milton Park, Abingdon, Oxon, OX14 4RN
and by Routledge
605 Third Avenue, New York, NY 10017

Routledge is an imprint of the Taylor & Francis Group, an informa business

© 1984 Istituto Affari Internazionali

Publisher's Note
The publisher has gone to great lengths to ensure the quality of this reprint but points out that some imperfections in the original copies may be apparent.

Disclaimer
The publisher has made every effort to trace copyright holders and welcomes correspondence from those they have been unable to contact.

A Library of Congress record exists under LCCN: 83040056

ISBN 13: 978-0-367-69841-6 (hbk)
ISBN 13: 978-1-003-14351-2 (ebk)
ISBN 13: 978-0-367-69836-2 (pbk)

The Mediterranean Region

ECONOMIC INTERDEPENDANCE AND THE
FUTURE OF SOCIETY

Edited by Giacomo Luciani

CROOM HELM
London & Canberra
ST. MARTIN'S PRESS
New York

©1984 Istituto Affari Internazionali
Croom Helm Ltd, Provident House, Burrell Row,
Beckenham, Kent BR3 1AT

Croom Helm Australia Pty Ltd, 28 Kembla St.
Fyshwick, ACT 2609, Australia

British Library Cataloguing in Publication Data

The Mediterranean region
   1. Mediterranean region
   I. Luciani, Giacomo
   909'.098220828   DE100

   ISBN 0-7099-1656-6

Library of Congress Cataloging in Publication Data
Main entry under title:

The Mediterranean region.

   "This book is the final product undertaken by the Istituto Affari
Internazionali. At the end of a one-year research phase . . . a conference was
held at Villa Montecucco in Castelgandolfo on September 6-9, 1983" – P.
   Includes bibliographical references and index.
   I. Mediterranean Region – Foreign relations –
1945-   – Congresses. 2. Mediterranean Region –
Strategic aspects – Congresses. 3. Natural resources –
Mediterranean Region – Congresses. – 4. Maritime law –
Mediterranean Region – Congresses. 5. Mediterranean
Region – Politics and government – 1945-   – Congresses.
I. Luciani, Giacomo, 1948-   .   II. Istituto affari
internazionali.
DE100.M43   1984   327'.09182'2   83-40056
ISBN 0-312-52818-3 (St. Martin's Press)

CONTENTS

List of Tables   i
List of Figures   iii
Note on contributors   iv
Acknowledgements   v
Introduction   vii

PART ONE: ENERGY AND OTHER RESOURCES

1.   THE MEDITERRANEAN AND THE ENERGY PICTURE
     by Giacomo Luciani   1

     Introduction   1
     Hydrocarbon Exploration and Production   1
        The nature and configuration of the
        Mediterranean seabed   1
        The development of offshore technology   2
        Current offshore production   6
        Offshore exploration   8
     Oil Transportation   15
        Determinants of Mediterranean oil traffic   15
        The redirection of oil logistics   15
        Developments in Iraq in the 1970s   17
        The Saudi East-West pipeline   19
        The Sumed and the reopening of the Suez Canal   20
        New strategic pipelines under construction   22
        Future evolution of oil traffic   23
     The Mediterranean and the Development of Natural
     Gas Resources   24
        Some peculiarities of natural gas   24

Current infrastructure for gas trade   26
Possible developments of gas trade   31
Offshore gas reserves   36
Conclusion   37
Notes   39

2. MEDITERRANEAN NON-ENERGY RESOURCES: SCOPE FOR
    COOPERATION AND DANGERS OF CONFLICT.
    by Gerald H. Blake   41

Introduction   41
Fishing   47
  Importance of Mediterranean fisheries   47
  Exclusive fishing claims and disputes   49
  Conclusion - Fishing   53
Seabed Mining   55
  Introduction   55
  Mediterranean seabed resources   56
  Conclusion - Seabed mining   60
Environmental Protection   61
  Introduction   61
  Types of pollution   62
  International cooperation to combat
  Mediterranean pollution   65
  Conclusion - Environmental protection   67
Conclusion   68
  Fishing   70
  Seabed mining   70
  Environmental protection   71
Notes   72

PART TWO: LEGAL CONSIDERATIONS: TERRITORIAL LIMITS,
          CONTINENTAL SHELF AND THE LAW OF THE SEA

3. EXTENSION AND DELIMITATION OF NATIONAL SEA
    BOUNDARIES IN THE MEDITERRANEAN
    by Geoffrey Marston   75

The Physical Background   75
The Legal Background   77
  The International Law of the Sea: the 1958
  Convention   77

Relevant concepts in the International Law
of the Sea   78
The International Law of the Sea: the work of
UNCLOS III   86
The International Law of the Sea: the work of
UNCLOS III in developing new concepts   96
The legal status of the 1982 Convention on the
Law of the Sea   102
The practice of the Mediterranean states   104
Cooperation between Mediterranean States   113
The Legal Aspects of Future Delimitations
in the Mediterranean   117
Notes   125

4. THE ARAB STATES AND MAJOR SEA ISSUES
   by Nazih N.M. Ayubi   126

Arab States and the Third UNCLOS   126
General Attitudes and Positions   128
Cooperation and Conflict in the Mediterranean
Sea   130
The Red Sea: An Arab Lake?   134
The Persian Gulf: A Sea of Crisis   136
The Issue of Foreign Military Bases   139
Conclusion   141
Notes   143

PART THREE: ASPECTS OF POLITICAL AND MILITARY CONFLICT

5. MAGHREBI POLITICS AND MEDITERRANEAN IMPLICATIONS
   by I. William Zartmann   149

Introduction   149
General Sources of Conflict in Maghrebi
Politics   151
   Borders   151
   National consolidation processes   154
   Ideology   156
   The checkerboard pattern of relations   159
   Position within regional contexts   161
   Position in global politics   163

The Sea as a Source of Conflict   166
  Boundaries   167
  Bridges across the sea   169
  The sea as a potential battlefield   169
  Bases and external military presence   170
  Perceptions of the geostrategic significance
  of the Mediterranean   172
The Sea as a Door through which Outside
Conflicts Penetrate the Region   174
  Relations with the superpowers   174
  Breakdown of regional organizations   174
  Superpower hegemony and conflict resolution   175
  The potential for Soviet penetration   175
  Two models for the future   176
Notes   177

6. SUBNATIONAL CONFLICT IN THE MEDITERRANEAN REGION
   by Brian M. Jenkins   179

   Patterns of Conflict in the Mediterranean   179
   The Cradle of International Terrorism   181
   A Labyrinth of Secret Wars and Secret Deals   183
   Terrorism in the Maritime Environment   186
   Are Maritime Targets Attractive to Terrorists?   187
   Current and Future Capabilities   190
   Prospects for Cooperation   192
   The Outlook   194
   A Chronology of Conflict in the Mediterranean   196
   Notes   203

7. THE MILITARY PRESENCE OF THE RIPARIAN COUNTRIES
   by Maurizio Cremasco   206

   The Frame of Reference   206
   A Summary Description of the Current Situation   213
   The Implications of a New Situation   219
   Notes   227

PART FOUR: EUROPE, THE SUPERPOWERS AND THE
            MEDITERRANEAN

8. EUROPEAN CONCEPTS FOR THE MEDITERRANEAN REGION
    by Elfriede Regelsberger and Wolfgang Wessels   239

    Introduction: The Political Challenges   239
    The Historical Devlopment of an EC
    Mediterranean Policy   240
      The first phase of Community policies   240
      European Political Cooperation and the
      Mediterranean region   242
      The Euro-Arab dialogue and concepts for
      group-to-group negotiations   243
      Towards a new phase?   244
    The Importance of the Mediterranean Region
    for the EC   246
      General political and security interests   246
      Cultural-historical links   248
      The management of common Mediterranean problems 249
      Some conclusions   249
      The EC internal dimension   251
    Global Options   251
      First set of options: a "mare nostrum" of the
      littorals   251
      Second set of options: diversifying EC policies 257
    A Need for Decisions?   260
    Notes   262

9. SOVIET STRATEGY AND THE OBJECTIVES OF THEIR NAVAL
    PRESENCE IN THE MEDITERRANEAN
    by Robert G. Weinland   267

    Introduction   267
    Naval Strategy   268
    Planning for Wartime   269
    Policy in Peacetime   274
      Active defence of peace and progress   276
      Defence of peace   276
      Defence of progress   280
      Preparation of maritime theatres of military
      operations   282

The Evolution of the Soviet Naval Presence in
the Mediterranean   284
The Future   287
Notes   290

10. AMERICAN FOREIGN POLICY, NATO IN THE
    MEDITERRANEAN AND THE DEFENCE OF THE GULF.
    by Ciro Elliott Zoppo   292

    East-West Security in North-South Politics:
    A Crucible for Interdependence   292
    The Defence of the Gulf and Mediterranean
    Security   295
    Developments in Military Technology and the Chan-
    ging Character of Security in the Mediterranean   298
    Nato Aspects of Gulf Contingencies and Superpower
    Rules of Engagement in the Mediterranean   302
    U.S. National Interests Linking the Mediterranean
    to the Gulf as a Function of East-West Factors   311
    Issues and Prospects for Security   316
    Notes   323

Index   325

LIST OF TABLES

1.1   Offshore daily average crude production   6
1.2   Average daily production and discovery date
      of off-shore fields in the Mediterranean   7
1.3   Number of wells drilled per year in country   10
1.4   Production of crude oil in selected
      Mediterranean countries   16
1.5   Estimated spot tanker freight costs for Arabian
      Light to representative destinations from either
      Yanbu or Ras Tanura, various transportation
      alternatives   21
1.6   Estimate of Mediterranean crude oil shipments
      in 1985   24
1.7   Natural gas reserves around the Mediterranean 27
1.8   Operational L.N.G. projects at the end of 1980 30
1.9   Cost comparison of alternative modes of
      transporting natural gas   34-35
2.1   The importance of Mediterranean coastlines
      to Mediterranean states   44
2.2   Allocation of Mediterranean seabed according
      to hypothetical boundaries in figure 2.1   45
2.3   Mediterranean fish catches by country   46
2.4   Average limit value of fish landings in 1970   48
2.5   The most important species by value in
      Mediterranean and Black Sea catches in 1977   50
2.6   Mediterranean Sea: territorial water claims
      and exclusive exclusive fishery claims   51
2.7   Fish catches in the eastern Mediterranean
      in 1978   53
2.8   Fish and fish products by value in 1979   54

2.9   Sources of oil pollution in the world's oceans   64
2.10 Suez Canal: north-bound petroleum and products
      by unloading countries 1981   69
7.1   Spread of anti-ship missiles in the
      Mediterranean   229
7.2   Characteristics of surface-to-surface
      missiles deployed in the Mediterranean
      area   230-231
7.3   Characteristics of air-to-surface missiles
      deployed in the Mediterranean area   232-233
7.4   Corvettes, fast attack craft, hydrofoils
      armed with surface-to-surface missiles   234-235
7.5   Characteristics of selected combat aircraft
      deployed in the Mediterranean area   236
7.6   The Mediterranean naval market 1970-1981   237-238

LIST OF FIGURES

1.1   Depths profile of the Mediterranean Basin   4-5
1.2   Oil finds in the Mediterranean offshore   9
1.3   Areas opened to exploration in the Mediterranean
      offshore   11
1.4   Transmediterranean pipeline   28
1.5   Strait of Messina and Sicily Channel crossings   29
1.6   Possible gas exports to western Europe
      from Algeria and Arabian Gulf   32
2.1   Maritime boundary status (1982)   43
2.2   Submarine volcanoes   59
3.1   Delimitation of EEZs between Italy and Spain   108
3.2   Delimitation of EEZs between Italy and Tunisia   109
3.3   Delimitation of EEZs between Italy and Greece   110
3.4   Delimitation of EEZs between Italy and
      Yugoslavia   111
9.1   Soviet Naval Presence in the Mediterranean:
      1964-1981
9.2   Soviet Naval Presence Worldwide: 1964-1981

Nazih N.M. Ayubi is a professor of Political Science at the University of Exeter, U.K.

Gerald H. Blake is a professor of Geography at the University of Durham, U.K.

Maurizio Cremasco is a consultant on Strategic Affairs at the Istituto Affari Internazionali, Rome.

Brian M. Jenkins is the director of the Security and Subnational Conflicts Programme at the Rand Corporation, Santa Monica, California.

Giacomo Luciani is the director of studies at the Istituto Affari Internazionali, Rome.

Geoffrey Marston is a fellow of the Sudney Sussex College, Cambridge, U.K.

Elfriede Regelsberger is a researcher at the Institut fuer Europaeische Politik, Bonn.

Robert G. Weinland is an expert in Strategic Affairs at the Center for Naval Analysis, Alexandria, Virginia.

Wolfgang Wessels is the director of the Institut fuer Europaeische Politik, Bonn.

I. William Zartmann is the director of African Studies at the School of Advanced International Studies, Johns Hopkins University, Washington, D.C.

Ciro Elliott Zoppo is a professor of Political Science at the University of California, Los Angeles.

ACKNOWLEDGEMENTS

This book is the final product of a research project undertaken by the Istituto Affari Internazionali with funds provided by the Ford Foundation under grant no. 810-0796. At the end of a one-year research phase, an international conference was held in the Villa Montecucco, in Castelgandolfo on September 6-9, 1983; the conference was made possible by a grant from the Dipartimento della Cooperazione allo Sviluppo of the Italian Ministry of Foreign Affairs, and a grant from the ENI (Ente Nazionale Idrocarburi). The Institute is grateful for this very essential support.

The Conference in Castelgandolfo was attended by participants from 18 countries, including most Mediterranean riparian countries and the United States and Arab and European countries which do not border the Sea. At the end of the conference, participants were received at the Quirinale Palace by the President of the Republic of Italy, Sandro Pertini.

The discussion at the Conference greatly helped all authors in refining the first drafts of their papers. Other papers were submitted to the conference which are not included in this book: they were published in Lo Spettatore Internazionale, IAI's quarterly journal in English, no. 4, 1982.

I wish to acknowledge the very useful assistance which I received at different stages from various staff members of the Ford Foundation, and particularly Enid Schoettle, Ann Lesch, Peter Ruof and Gary Sick. I wish to thank Dr Luigi Cavalchini, Minister Plenipotentiary, for the attention with which he followed this undertaking and the effective help he gave in securing some of the funding. Dr Domenico Tantillo and Dr Gilberto Gabrielli, of ENI, supported our work on this occasion, as they constantly do with IAI activities.

In designing the project and editing this book I very greatly benefited from the suggestions and criticisms of Prof. P.J. Vatikiotis.

The IAI has focussed on Mediterranean problems for more than a decade. This project is one of a series on problems related to the region. More specifically, the immediate intellectual origin of this project is to be found in another Conference organized by the Center for International and Strategic Affairs of the University of California Los Angeles. ("The Mediterranean in World Politics", Dec. 1980, Castelgandolfo).

During the initial phase of the present undertaking I greatly benefited from a stay in Los Angeles as a visitor at CISA, and with partial support from CISA itself: this allowed me to discuss the contents of their repective papers with Nazih Ayubi, Brian Jenkins and Ciro Zoppo, in person. For this I am grateful to Prof. Roman Kolkowitcz, who was the Director of CISA at the time of my visit, and to Prof. Michael Intriligator, who has since succeeded him. My debt to Ciro Zoppo is not just for the chapter that he wrote, but for the numerous useful suggestions that he offered at all stages, and the friendly help in Los Angeles. For the latter I am also indebted to Rosemary Zoppo.

Formal editing of the papers and preparation of the index were accomplished by Diane Cherney. The camera copy was produced by Rosanna Hatalak and Elizabeth Williams.

My work was always supported morally and materially by all my colleagues at IAI, and the merit, if any, of the realization of this project is not personally mine, but belongs collectively to the Institute.

As usual, all errors or shortcomings are my full responsibility.

G.L.

INTRODUCTION

by Giacomo Luciani

The Mediterranean Sea and the region surrounding it have consistently played a crucially important role in international relations since the end of World War II. Mostly because of the presence of multiple conflicts, the Mediterranean has attracted the attention and presence of both the United States and the Soviet Union; some believe that this presence is in itself a major source of conflict within the region.

The focus on the Mediterranean dimension does not need justification from an historical point of view. Although the Mediterranean has not been politically united since the division of the Roman Empire, the shifting boundaries of successive empires and other political entities overlapped in such a way that all Mediterranean countries share at least parts of a common historical heritage. Since the dismemberment of the Roman Empire a region like Sicily has been successively Greek, Arab, Norman, French, Spanish, part of the Kingdom of the Two Sicilies, and finally Italian. The Arab tide engulfed almost the entire Iberian peninsula, while the Ottoman empire controlled almost all the Balkans. The Italian merchant cities controlled a number of trading stations along the Asian coast, while Spanish presidia dotted the northern African littoral. The French, and to a less extent the Italian, colonial conquests contributed a further piece of common Mediterranean history. Greek temples, Roman amphitheatres, Egyptian obelisks, Arab and Ottoman mosques, fortresses and castles built by Crusaders and Venetians, are spread out in the region and remind the traveller of the shifting crosscurrents of Mediterranean history.

The outcome of these multiple historical currents is, on the one hand, a large degree of commonality in cultural background, and, on the other, a predominance of segmentation and conflict at the political level. Thus, while the idea of a Mediterranean community has always been present and lively in the area, it is extremely dubious that the Mediterranean as such constitutes a meaningful subject for political and economic analysis.

The reason why this book does focus on the Mediterranean as the dimension for analysis is that it tries to assess the impact of certain new developments specifically relating to the sea, rather than to the region surrounding it. The sea itself is often subdivided into sub-basins: a Western Mediterranean is often distinguished from an Eastern Mediterranean (sometimes: a Central and an Eastern Mediterranean) while the Adriatic and the Aegean are usually discussed separately. While these subdivisions are obviously legitimate, it is appropriate in this book to focus on the entire basin because the factors under analysis are equally at work in all its constituent parts. They clearly do not have the same sign and intensity in each constituent part, but they are present throughout the region.

The basic hypothesis which this book is intended to explore is that some recent developments are creating, on the one hand, greater occasions and incentives for cooperation among Mediterranean countries and, on the other, fresh danger of conflicts. In short, developments specifically related to the role of the sea are intensifying the links of interdependence among Mediterranean countries, an interdependence which may just as well breed cooperation as conflict.

Clearly, these developments are in no way unique to the Mediterranean, but rather reflect global trends. Thus the growing (or perhaps simply continuing) importance of access to natural resources as a motivation in international relations is a characterizing feature of the last decade, while the juridical treatment of maritime activities is used as a case study in one of the reference works on interdependence in international relations (Power and Interdependence, by Keohane and Nye).

However, conditions in the Mediterranean are considerably more complex and difficult than they are in most other basins. Complexity is the result of the high number of independent riparian countries, the numerous conflicts between them, the simultaneous presence of the top super-powers' naval forces, the proximity to key areas of super-power confrontation, namely central Europe and the Gulf; it is also a consequence of the shape of the basin, the presence of numerous peninsulae and islands and the conform-ation of the seabed. Further one may add the continuing importance of the Mediterranean to international maritime traffic and the polarization among riparian countries in terms of international allegiances. Not suprisingly, this generates an insoluble Gordian knot, in truly Mediterranean tradition.

*   *   *

But then, what is new? The book points to three main elements of change, which play, so to speak, the role of independent variables. These are: 1) the riparian countries' growing economic interest in gaining control over the sea; 2) the evolving international legal regime concerning the seas; 3) the technological progress in weapons systems, and the accelerating diffusion of advanced weapons systems among Mediterranean riparian countries.

The first independent variable is discussed in Chapters One and Two of this book. The economic interests involved are numerous, and a distinction between energy related and non-energy related issues has been made. Among the former, the Mediterranean is a very important route for international oil trade; and the Mediterranean basin may conceal larger hydrocarbon resources than those that have been discovered so far. Among the latter, the discussion covers seabed minerals, fisheries, and preservation of the environment.

It is interesting to note that although both discuss-ions conclude that the Mediterranean basin is potentially important for the economic prosperity of the riparian count-ries, they also suggest that the development of resources to be found in the sea ought not to be controversial. This is partly for economic reasons (e.g. current international prices do not warrant the mining of seabed minerals); partly because of technological factors (e.g. oil production cannot presently be carried out under more than 1,000 feet of water, and most of the Mediterranean seabed lies deeper than that); or finally for geographic reasons, because most of the promising areas happen to lie in waters which are not controversial, pertaining clearly to only one of the riparian countries. The main exception to this statement is the area lying between Sicily, Tunisia, Libya and Malta.

Conflict may be more acute for mobile resources, such as fish or pollutants (the latter may be viewed as a negative resource), but in this respect some, if not a great, degree of international cooperation among riparian countries already exists, in the form of the Barcelona Convention and subsequent documents and initiatives (the Mediterranean Action Plan).

One would therefore expect that a favourable climate would be perceived to exist for cooperation among the Mediterranean countries in discovering and developing what valuable resources there are at sea. However, the contrary appears to be true: the fact that the prospect for economic utilization lies rather far in the future, that the economic value of resources present in the Mediterranean cannot be assessed in any meaningful way, and that there is little to be gained from conflict - most of the known resources being non-controversial - does not seem to favour cooperation. Not only is it not possible to launch a pan-Mediterranean cooperation scheme, but a vast majority of the needed bilateral treaties delimiting each country's Exclusive Economic Zone (EEZ) remain to be negotiated. Conflict is active not only in those sections of the sea where indeed economic interest may be a plausible explanation, such as in the waters off Sicily, Tunisia, Libya and Malta, but also in the Aegean, whose prospects for mineral deposits are almost unanimously rated as poor. Indeed, while there has been conflict in the former area, delimitation is progressing: an agreement exists between Italy and Tunisia; the latter and Libya seem inclined to accept the delimitation between them that was proposed by the International Court of Justice, and a similar procedure may be followed by Libya and Malta. Finally,a working arrangement exists between Italy and Malta.

Cooperation on environmental issues exists, and ought not to be discarded as irrelevant; at the same time its limitations are very easily seen. Because pollution originates predominantly in the industrial countries of the

Northern shore, the agreements that have been reached up to now bear practical consequences for only a very few of the riparian countries. These countries are basically reflecting their own concern for the environment, and cannot be said to respond to pressures coming from the non-industrial countries. The latter are mainly concerned to see that there is no negative consequence for their industrialization effort, i.e. are there to protect their right to pollute. Whenever the industrial countries are not ready to sacrifice their interests for the defence of the environment, no agreement is reached. Thus cooperation on the environment provides a favourable framework to at least discuss further issues, but seems unlikely to be a harbinger of future schemes with wider impact.

*   *   *

Partly because the economic interest in maritime resources has increased everywhere in the world, the legal regime for maritime activities is rapidly evolving. The lack of agreement at the global level was clearly shown by the concluding phases of the Third United Nations Conference on the Law of the Seas. The passage from the Carter to the Reagan administration in the United States brought a dramatic shift in the American position on the subject, leading to refusal to sign the text that had been negotiated within the U.N. Conference, and thereby throwing a substantial shadow on the future practical relevance of it. The surfacing of doubts among the same industrial lobbies which had staunchly opposed the U.N. draft raises the possibility that the dramatic shift that occurred may not be the last one. In any case, as is clearly underlined in Chapter Three of this book, many of the provisions of the U.N. text may be considered as being already valid, having been established by previous conventions, by practice or by implicit international acceptance.

What matters most to our argument is that both the U.N. text and other rules which may be considered to be internationally accepted do not easily apply in the objective conditions of the Mediterranean basin. It is not by chance that the long history of the U.N. Conference was initiated by the representative of a Mediterranean country,

Ambassador Pardo of Malta, proposing an internationalistic approach, i.e. a declaration to the effect that the sea belongs to humanity as a whole. Almost nothing of this original inspiration is left in the final text, which contains quite to the contrary detailed rules indicating how the sea may be subdivided among individual countries. This nationalistic approach is bound to multiply conflicts in an environment such as the Mediterranean one, where countries step on each other's toes as soon as they get into the water!

Given geographic realities, the process of defining boundaries for national internal waters, territorial waters and exclusive economic zones is bound to generate conflict. It is also bound, legal assurances notwithstanding, to generate apprehension over the preservation of effective freedom of navigation.

The difficulties that may derive from the definition of internal waters, especially in connection with the delimitation of "historic bays" are highlighted by recurrent incidents between the Sixth Fleet and Libyan planes above the Gulf of Sirte. However, the Libyan claim is not the only example of strained utilization of the concept of historic bay, which Italy, for example, utilizes to define as internal waters the entire Gulf of Taranto. The existence of archipelagos generates conflict not only, as is well known, between Greece and Turkey, but between France and Italy as well, concerning definition of base lines in the northern Tyrrhenian Sea, between Corsica and Tuscany. An important implication of this, which also emerges from the analysis of the position of the Arab countries in Chapter Four, is that conflicts over delimitation of national maritime interests is not exclusively or primarily a North-South affair.

Politically speaking, the work of the U.N. Conference was seen as a compromise between actors whose main interest lies in freedom of navigation and parties that wish to establish exclusive national control over maritime resources. From this point of view, one might claim that the evolving legal regime may at least have the advantage of more clearly guaranteeing freedom of navigation, which undoubtedly is the main economic, political and security concern of each Mediterranean country. However, in practice, the need to protect existing or potential economic interests at sea can be expected to lead to fast growing projection over the waters of the riparian countries' military presence.

It is in this context that consideration of terrorism plays a key role. The inclusion of a chapter on this item (Chapter Six) does not imply that the Mediterranean countries and peoples are for some reason inclined or lenient towards terrorism; at the same time there is little doubt that a large number of terrorist events have taken place in the Mediterranean countries, that a large number of organized movements resort to terrorist practices there, and that structural conditions (referring both to geographic structure and to the socio-political environment) are strongly favourable to the adoption of this method of political struggle.

The multiplication of important economic targets at sea can only increase the occasions for terrorist activity. Both moving and stationary objectives offer good opportunities for the accomplishment of resounding actions of the type which terrorists pursue. But then the implicit, if not immediate, existence of a terrorist threat to economic interests at sea could be taken as sufficient justification on the part of the riparian governments to substantially increase their ability to project force at sea.

Speaking more generally, it seems unlikely that a total divorce be possible between the economic and security facets of the sea. Legally speaking, riparian countries will enjoy exclusive control over resources in their EEZs, while at the same time naval forces will be able to roam around freely, without any aggressive implication, provided they do not enter the territorial waters of a foreign country. In practical life, however, one expects that if riparian countries add their forces to the already substantial ones stationed in the Mediterranean by the United States and the Soviet Union, and forces from all sides start to roam freely into other countries' EEZs where highly sensitive economic equipment might be present, dangers of conflict or tensions will inevitably be increased.

Thus, there is no need to presuppose an outburst of terrorism at sea, nor indeed is it necessary that important economic installations are multiplied in the EEZs: the tendency to rapid and indeed accelerating growth in the riparian countries' capability to project force at sea is already with us.

*   *   *

Thus we come to the consideration of the third "independent variable" that we mentioned above: technological change in weapons systems. Indeed, the growing capabilities of the riparian countries which are analysed in Chapter Seven are only partly a consequence of quantitative growth in the weapons systems they have at hand, and mostly a consequence of the improved characteristics of these systems. The growing importance of electronics carries substantial consequences with regard to the identification of targets and the accuracy of offensive systems. In the maritime environment the rules of the game have changed markedly as a consequence, and size is no longer such an overriding factor. Thus even if no riparian country can even compare with the forces which the superpowers keep stationed in the Mediterranean, the kind of systems which are now available to them are capable of inflicting not irrelevant damage.

A second crucial technological factor is the improvement in the range of land-based aircraft, which allows riparian countries to cover a much larger portion of the sea with their air forces. In this respect the Mediterranean has become not just narrow, as indeed it is, but also shorter than it used to be: Soviet Backfire bombers stationed in Crimea can cover objectives in the western Mediterranean as well as in the eastern portion of it.

Thus, even if no consideration is given to the growth of economic interests and the potential restrictions that may derive from it, to the increasing complexities of the internationally accepted legal regime and the conflicts that they may lead to, strategic realities are evolving in a direction that changes the very nature of the Mediterranean environment. The Mediterranean is narrowing, shares less and less of the nature of the High Seas, and increasingly appears as not much more than a large fjord or a complex system of straits.

While this may have no immediate or automatic consequences on the perceptions and actions of all the parties involved, it is a transformation which is bound to affect, in particular, the posture of the superpowers and their relations with their respective allies among the riparian countries. To the naval forces of the superpowers the

Mediterranean now presents some risk of interference from the riparian countries, and their presence in the EEZs of non-allied countries may acquire an unintended implication of threat or interference. In other words, the Mediterranean is becoming a less convenient base for the conduct of naval diplomacy in peacetime.

* * *

Our belief that the Mediterranean maritime environment is changing does not however imply that serious and autonomous conflict could erupt as a consequence. Although disagreement over delimitation of internal or territorial waters, or delimitation of EEZs, or exploitation of existing resources may be difficult to overcome and may lead to further tensions in relations between Mediterranean countries, the likelihood that such tension will per se result in armed conflict can be rated as practically nil.

There will not be a Falklands war in the Mediterranean, although much was said about the existence of huge mineral resources in the continental shelf which pertains to those islands. That war was fought by both sides more in view of internal political benefit than in order to achieve control over hidden resources. It was fought between two distant adversaries, who have little or no political intercourse at other levels. Distance allowed the political containment of the conflict, and a military solution to it. No such conflict is feasible in the Mediterranean because countries are too close to each other, they are too deeply interrelated, either directly or indirectly, and conflict over maritime issues would not remain confined to those issues alone.

Indeed, Mediterranean countries are engaged in more than one armed conflict between them, on issues which are much more immediately important to them than control of maritime resources. This inevitably leads to the consequence that much of their attitude towards maritime problems is a mere reflection of their overall interests and foreign policies. This point is underlined in Chapters Five and Eight, covering respectively the Maghreb and the European Community. These chapeters analyse the possible impact, in the one case, of a set of conflictual relations and, in the other, of a broad-based effort at regional integration. Not that the Maghreb is the most conflictual region around the Mediterranean: we could have obviously looked at the eastern Mediterranean instead, but it was felt that the magnitude of the conflicts found there would have made the question of the importance of maritime issues almost trivial. And as far as Greece and Turkey are concerned, the Community might a priori be expected to play a positive role in the resolution of conflict between one of its members and another country that has been tied to it by a treaty of association explicitly spelling out the objective of future membership.

The analysis does not lead to a clear conclusion. On the one hand in the case of the Maghreb, it appears that conflict over maritime issues is indeed little more than a reflection of a more general pattern of conflict, essentially motivated by domestic political mechanisms and needs. On the other, the record of the European Community is also rather negative, as it does not appear to be able to play a significant role in the resolution of maritime conflict between its member countries, or with closely associated countries such as Turkey. In other words, the existence of a broader framework of economic integration and cooperation is not necessarily a key to the resolution of conflict related to maritime problems.

But is it so because these conflicts are perceived to be so important as to make compromise difficult even in the context of a multi-faceted relationship? Or is it simply a reflection of the Community's creeping crisis, a manifestation of the declining drive to push forward the process of European integration? While one might be more inclined to opt for the second, and believe that a more vigorous and healthy climate in European affairs would make it quite easy to engineer the communal approach needed to overcome conflicts, there is little point in speculating on what might happen if the Community were not in disarray, as it so deeply is.

Although the Community has tried to project itself into international affairs as a "civilian power", with a role clearly different from that of the superpowers, the apparent impotence regarding the question of exploitation of Mediterranean resources - a typically civilian issue - bodes ill for the possibility of successfully mediating more difficult tangles with a predominantly security characterization, such as the Arab-Israeli conflict. Thus, alas, it is not from Brussels that we may expect a fresh approach to the problems of interdependence in the Mediterranean basin.

*   *   *

Where the new developments that we underline will make a difference is primarily in relations between the super-powers and their allies in the region. The problems that both the Soviet Union and the United States are facing are discussed in detail in the final two chapters of the book.

In both cases, although in different ways, the pre-sence and role of the naval forces which are normally stationed in the Mediterranean basin is under debate. Economic, legal and technological developments are gradually limiting the flexibility and increasing the vulnerability of those forces.

In the past, the naval presence of the Sixth Fleet in the Mediterranean did not greatly depend on the cooperation of Mediterranean NATO allies and other friends of the United States, while it allowed the latter to express significant support for one or the other of the Mediterranean countries, according to the interests of American foreign policy. While in essence this may still be true, today the continued presence of the Sixth needs more significant support on the part of the Mediterranean allies, and her effectiveness in emergency conditions, when she is needed most, is much more dubious. Politically speaking, direct naval presence is less and less an alternative to the need for the United States to consult and coordinate with its allies in the region.

On the one hand, it is clear that this is no argument to reduce or eliminate the naval presence: coordination with the allies would still be necessary, and one could easily argue that it would become even more problematic. On the other, the politically important point is that a set of concurrent factors increases the need for active cooperation from regional allies.

The recent behaviour of the Soviet Eskadra is an implicit confirmation of this conclusion. Although the Russian fleet is still there, the telling fact is that it has proved totally ineffective in emergencies. Soviet ships did not interfere with the Israeli siege of Beirut in the summer of 1982, nor does one hear of close encounters with the Sixth Fleet. As is suggested in Chapter Nine, this is probably related in part to the loss of Egypt as an important regional ally. Deprived of the effective support of regional allies, the Eskadra cannot afford to be as visible as it sometimes was in the past.

*　　*　　*

Some of the countries that border on the Mediterranean have called for a substantial or complete reduction of the military presence of the superpowers in the Mediterranean basin, as a first step to a peaceful and cooperative approach to Mediterranean problems. This idea, in the words of Charles Zorgbibe, of La Méditerrannée sans les grands, would solve the problem of defining a division of labour between the superpowers and their local allies by simply eliminating the former. It would, no doubt, be a neat solution to the problems posed by growing interdependence and the progressive shrinking of the basin.

Unfortunately, it is based on premises which appear to be totally unrealistic. Conflicts between Mediterranean countries are sometimes fuelled by the outside powers, but they are generated within the region. The presence of the superpowers may encourage irresponsible behaviour on the part of the contendents, and it may delay a solution almost ad infinitum. Yet, there is indeed little evidence that all that is needed to open the way to a peaceful and cooperative Mediterranean is to remove the presence of the superpowers.

Besides, superpower presence does not necessarily occur in direct form only: it could be realized by proxy. Simple elimination of direct military presence would not eliminate interference, certainly not if arms deliveries were not somehow curbed as well. But then, assuming for a moment that it were possible to agree that neither the Soviet Union nor the United States would supply weapons to any Mediterranean country, this would simply lead to an unacceptable imbalance between countries that have the capability to produce weapons domestically and those that do not.

The reasoning ought to be reversed: a reduction in the role and presence of the superpowers is the last, not the

first step along the road to Mediterranean cooperation. It
is the outcome and conclusion, not the premise. It will be
achieved only if and when countries in the region prove that
they are able to manage their interdependence and solve
their quarrels with different instruments than the use of
armed force.

The growing role which the local powers should play in
a new division of labour with the superpowers is no different. What is mostly requested of them is a patient and
continuous effort to defuse conflicts before they reach a
critical point. In this respect the maritime issues discussed in this book may be an important terrain for diplomatic activity.

*   *   *

The question of some form of arms limitation agreement
is also raised from time to time. In the summer of '83,
Malta delayed the conclusion of the Madrid Conference on
European Security and Cooperation with the demand that a
similar Conference be called for the Mediterranean region, a
request which raised a lot of scepticism and little enthusiastic support. However, sooner or later a conference on
Cooperation and Security in the Mediterranean is likely to
be convened. What are the prospects for meaningful
discussions?

The main obstacle to an agreement for arms limitation
in the Mediterranean is in the nature of the region and the
sea. The Mediterranean region certainly has no independent
status, but is overlapped by bilateral and multilateral
alliances with non-Mediterranean actors. Thus there is
little point in discussing arms limitation in the Mediterranean region separately from arms limitation in central
Europe or in the Gulf. And there is little hope of achieving
significant results as long as there is open armed conflict,
as between Israel and its neighbours.

If we drop the notion of the Mediterranean region and
concentrate instead on the sea, the feasibility of an arms
limitation agreement is certainly improved, because of the
preponderance of superpower forces. However, the significance of any such agreement would necessarily be very
limited. By definition (because of the distinction between

the sea and its surrounding region) the potential agreement would apply to naval forces and sea-based forces only. However, as was underlined earlier, it is the land-based forces which are playing a growing role in controlling the seas. Even if it were possible, total elimination of sea-based weapon systems would not greatly change the level of confrontation from a military point of view.

At the same time, the fact that control restricted to sea-based systems would admittedly have limited significance does not mean that it would be entirely useless, nor that it is not worthwhile to discuss it. Quite to the contrary, this book underlines some of the reasons why it would still be important to achieve some agreement on limitation: the fast growth of naval systems that may otherwise be expected is in fact especially prone to generating conflict in connection with the growing interest over marine resources and the uncertainties of the international legal provisos. An increase in the presence of land- based systems, even if potentially employable over the waters, is felt to have less potential for generating incidents.

*　　*　　*

The peculiarity of this book is to have approached a complex subject from numerous different angles and disciplines. While the different chapters are conceived to fit together and substantiate the argument which we tried to explain in the preceding pages, they also have value per se, for the information and analysis which each contains. It is my hope that the reader will find this book useful also as a work of reference on the economic, political, legal and strategic problems of the Mediterranean Sea.

As one who finds great enjoyment in wandering over and around the Mediterranean, I also hope that this modest effort will contribute to a peaceful, sensible solution to the many conflicts that plague it.

Part One

ECONOMIC CONSIDERATIONS: Energy, other resources and conservation

Chapter One

THE MEDITERRANEAN AND THE ENERGY PICTURE

Giacomo Luciani

INTRODUCTION

This chapter describes energy-related developments in the Mediterranean Sea in order to assess, on the one hand, their importance in relation to the economic life of the riparian countries and the global energy picture, and, on the other, to provide a basis for the discussion of the interplay between economic, political and military factors.

The Mediterranean Sea comes into the energy picture in a variety of ways, not necessarily related to each other. The following discussion will concentrate first on the Mediterranean as a source of oil, that is, on the offshore exploration and production activities under way in the sea. In the second section, the issue of oil transportation will be considered, and the importance of the Mediterranean will be discussed in connection with developments in the broader Mediterranean region. In the third section, developments connected with the exploitation of natural gas resources will be discussed - an aspect in which issues of production, transportation and utilisation are inextricably interrelated. The final section is devoted to an attempt to draw a synthetic picture of relevance to the discussion of political and economic variables.

HYDROCARBON EXPLORATION AND PRODUCTION IN THE MEDITERRANEAN

The Nature and Configuration of the Mediterranean Seabed

Those who are accustomed to consulting political maps of the world or geographic maps showing elevations in detail but depicting the sea areas in a uniform shade of

1

blue may not realize that the Mediterranean is different from most other semi-enclosed seas in that it is very deep. From this point of view, the Mediterranean differs sharply from the North Sea, and also from the Red Sea which extends to great depths only in a narrow central section.

The African and European continents are moving closer together and "squeezing" the Mediterranean, causing the progressive sinking of the seabed which leaves both land masses with almost no continental shelves of noteworthy dimensions in the Mediterranean waters. In fact, almost 60% of the total seabed consists of deep sea plains between 6,000 and 9,000 feet below sea level (see Fig. 1.1).

There are only two sizeable areas where the continental shelf is considerably wide: in the Adriatic and in the zone between Sicily to the north, Tunisia to the west, western Libya to the south, and including Malta.

Elsewhere around the rim the continental shelf is generally very narrow and the seabed plunges almost immediately to depths of more than 3,000 feet. The most notable exceptions to this rule are the Nile Delta coast from Alexandria to El Arish, the Gulf of Iskenderun, the Gulf of Lions and the Spanish coast in front of the Baleares.

Finally, the seabed is very "corrugated" below both the Tyrrhenian and the Aegean Seas. The former is generally very deep - actually the deepest part of the Mediterranean - but contains a number of submerged mountains, mostly of volcanic origin (including active volcanoes). The Aegean is on average not as deep, but geologically it is highly fragmented.

The Development of Offshore Technology

Offshore hydrocarbon exploration and production is a dynamic industry in which technological development is progressing at a fast pace. It is only in the last fifteen years that the need to increase hydrocarbon reserves and diversify them geographically has greatly augmented the interest in exploration and production under increasingly difficult conditions.

Water depth is not the only obstacle that offshore exploration and production must overcome; others include the prevalent meteorological conditions, the degree of oxygenation of the water, the presence of ice, and so on. However, in the case of the Mediterranean, depth is the critical factor.

The Mediterranean and The Energy Picture

The horizon of maximum water depth for offshore activities is rapidly improving, and a sharp distinction must be
drawn between exploration and production.

In the case of exploration, there are drilling rigs
that are capable of operating in water depths of 6,000
feet. Exploratory wells are presently being drilled in
water depths of over 2,000 feet in numerous parts of the
world. In France, CFP-Total has joined forces with SNEA
(Elf) and the Institut Français des Pétroles in undertaking
a programme to improve deep sea technology. This led to the
development of the drillship Pélerin, which is capable of
drilling in water depths of more than 3,000 feet. The ship
completed her first well offshore Algeria (Habibas) in 925
metres of water, then was moved to the Atlantic. (1) Both
Agip and Eniepsa employ drillships considered capable of
drilling at similarly great depths, but which are currently
being used in shallower waters. (2)

On the other hand, the difficulty of carrying out production below great depths of water is much greater than for
exploration and the technological limit more stringent. The
difficulty of developing a field under water is two-sided:
there is the difficulty of laying pipes and the problem of
setting up a central production unit. The two constraints
are closely interrelated.

The record for laying pipes in deep waters has been
set by Saipem (ENI) with the Transmediterranean gas pipeline
which is covered by a maximum of 2,000 feet of water. The
same technology could be used for somewhat greater depths,
up to 3,000 feet.

All existing offshore production units are above the
water, that is, they are based on a platform of some kind. A
production platform, in contrast to a drilling rig, needs to
be permanently positioned above the field; it must, in other
words, sit on it. The tallest existing production platform
was installed by Shell over the Cognac field in Louisiana.
It operates in more than 1,000 ft. of water. However, its
cost is so high that it is believed that it represents the
economic limit beyond which an alternative technology
becomes convenient.

Such an alternative technology exists in the form of
underwater production systems. These, however, are not sufficiently developed yet and it is difficult to predict how
fast they will be able to push back the limit for hydrocarbon production in the deep seas. At present, a Shell-

Figure 1.1   Depths profile of the Mediterranean Basin

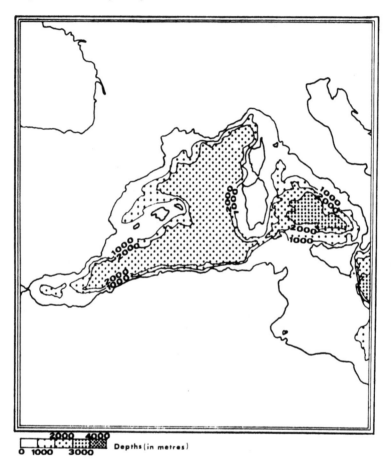

Depths (in metres)

Fig. 1.1/cont.

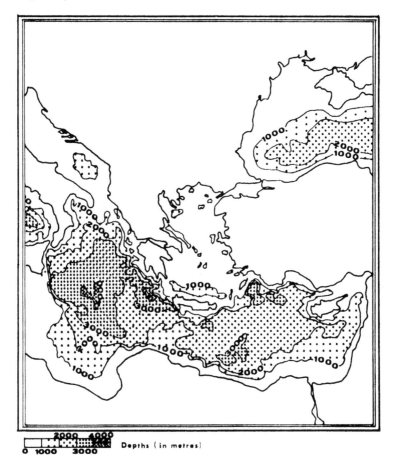

Depths (in metres)

Exxon partnership is installing underwater production systems for the Cormorant field in the North Sea, but in only 490 ft. of water - the companies claim, however, that "similar units can be used in waters several thousand feet deep". (3)

Therefore, while exploration is technologically possible almost everywhere in the Mediterranean, production is possible only in the limited parts that lie under no more than 1,000 ft. of water. This is, however, a limit that will certainly be substantially removed in the course of the next decade.

Current Offshore Production in the Mediterranean

Although production is logically a result of exploration, we will describe current offshore production in the Mediterranean before dealing with exploration to give a better impression of the present situation and the potential future developments.

Oil has been discovered in the Mediterranean off the shores of six riparian countries, namely Tunisia, Libya, Egypt, Greece, Italy and Spain. Both Spain and Egypt have a non-Mediterranean coast, and in the case of Egypt offshore discoveries in the Gulf of Suez are much more important than those in the Mediterranean. Discoveries off Libya and Egypt were made only in the 1980s and no production is yet under way. In Greece, offshore production started in 1981 and was taking place at a rate of around 25,000 b/d. (4) Table 1.1 gives the quantities of oil produced by the other three countries in 1975-1980.

Table 1.1: Offshore daily average crude production (,000 b/d)

	1980	1979	1978	1977	1976	1975
Tunisia	43.5	48.5	45.2	45.5	37.0	43.0
Spain	31.2	22.1	20.0	23.0	33.3	32.8
Italy	6.2	4.8	4.3	12.0	10.1	10.4

Source: Offshore, 20 June 1980 and 5 June 1981

6

These are small quantities, and they come from a small number of small fields. Table 1.2 shows production data broken down by individual fields.

The productive life of these fields has also proved to be very short. As an example, the Amposta field in Spain started production in 1973, produced around 1.9 million tons in 1974 and 1975, then declined rapidly to a level below 0.3 million tons in 1979 and 1980. (5)

Table 1.2: Average daily production (,000 b/d) and discovery date of offshore fields in the Mediterranean

Country and Field	Average Daily Production First Semester of		Date of Discovery
	1979	1980	
Tunisia			
Ashtart	45.3	43.5	1971
Spain			
Amposta Marina			
North	1.0	0.8	1970
Casablanca	n.a.	11.9	1975
Castellon B	4.8	7.6	1972
Dorado	7.4	2.7	1975
Italy			
Gela	8.1	10.0	1956
S. Maria "A" Mare	5.1	5.1	1974

Source: Offshore, 20 June 1980 and 5 June 1981

All the Spanish fields are located close to the continental coast in front of the Baleares. Of the two Italian fields, Gela is off the coast of southern Sicily while S. Maria is in the Adriatic. The new Greek field is in the northwest Aegean, not far from Thessaloniki.

To these producing fields must be added the new discoveries that have not yet been brought into production. These are particularly sizable in Tunisia, where four "exploitation concessions" were granted, in addition to the one covering the Ashtart field, namely: one to Total (Compagnie Française des Pétroles) covering the Isis field, one to SNEA (Elf) covering the Halk el Menzel field, and two to Shell for the Tazerka and Birsa fields. Fig. 1.2 shows the approximate location of these discoveries.

In Libya a field was found by Agip in an area called NC-41, lying northwest of Tripoli, some 75 miles off the coast; recoverable reserves there are estimated around 600-800 million barrels. Agip intends to develop this field at a cost of one billion dollars or more, and this would involve either offshore loading or the building of a pipeline to the Tripolitanian coast. (6)

The decision by the International Court of Justice in March 1982 left in Libyan waters a tract where two discoveries have been made by a group led by SNEA. (7)

In Egypt it was again Agip that discovered oil in the Mediterranean offshore, in the Gulf of al-Tina, about 30 miles east of Port Said. Indications are, however, that the oil-bearing stratum is not very deep.

Finally, Agip made a new discovery in 1980, in Italian waters 2,713 feet deep (Aquila field), yielding 3,000 b/d in test production. (8)

Offshore Exploration in the Mediterranean

Thus, current offshore production in the Mediterranan, although certainly not static, given that new discoveries are being made, is a negligible factor when compared to the global energy picture, and a marginal one when compared to the total consumption(in the case of Spain, Italy and Greece) or total production (in the case of Tunisia, Libya and Egypt) of the countries concerned.

This however does not give us an accurate feeling of the importance of the Mediterranean until we examine past and current exploration activity.

Offshore exploration activity, although growing, had been quite limited until the late 1970s, as is clearly shown by the data in Table 1.3.

Offshore exploration in the Mediterranean was perceived as not very attractive economically before 1970, except perhaps in Italy, to which every drop of oil was important for balance-of-payments reasons, almost independently of cost. More recently, obstacles of a technological nature, discussed above, as well political obstacles, have had to be overcome. The latter consist mostly in the lack of

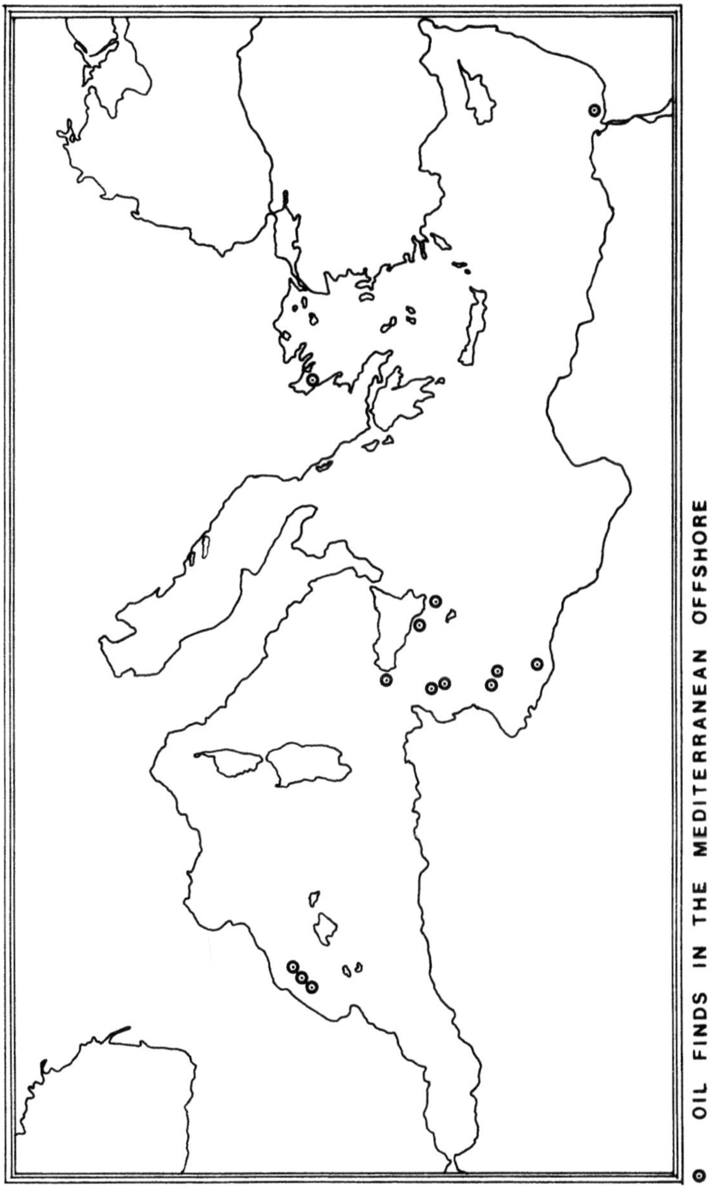

Figure 1.2 Oil finds in the Mediterranean offshore

OIL FINDS IN THE MEDITERRANEAN OFFSHORE

◉ OIL FINDS

Table 1.3: Number of wells drilled per year and country

	1979	1978	1977	1976	1975
Algeria	0	0	1	0	n.a.
Egypt*	37	34	44	33	21
France**	4	2	1	2	n.a.
Greece	6	8	n.a.	n.a.	1
Israel	0	0	0	1	0
Italy	36	37	27	22	n.a.
Libya	8	0	n.a.	8	n.a.
Malta	0	0	n.a.	n.a.	0
Morocco**	2	1	n.a.	n.a.	2
Spain**	17	19	17	25	11
Tunisia	16	9	n.a.	15	12
Turkey	0	1	0	4	1

*   includes wells drilled in the Gulf of Suez and the Red Sea
**   includes wells drilled in the Atlantic offshore

Source: Offshore, June 20, 1980; June 5, 1981

agreement between riparian countries on delimitations of seabed rights, as well as insufficient development of national legislation to regulate both the exploratory activities and actual production if oil is found. Yet exploration permits have been issued for large areas in the Mediterranean (Fig. 1.3).

No exploratory wells have been drilled in the Mediterranean offshore of Morocco, although Onarep, the Office National des Recherches Petrolières, holds an exploration permit and Amoco may also have requested one. Shell abandoned an exploration permit after carrying out seismic work.

The Algerian offshore is also almost unexplored, only a dry well having been drilled (Habibas).

The picture changes when we get to the Tunisian offshore, where almost the entire area is covered by various exploration permits. The lack of definition of a dividing line between Tunisia and Libya has been an obstacle, although not one that the companies have found entirely prohi-

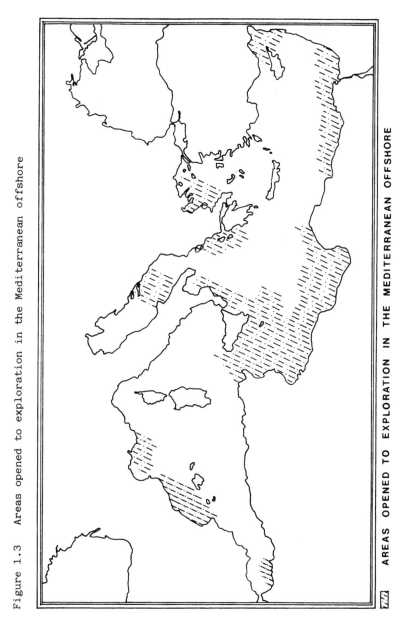

Figure 1.3   Areas opened to exploration in the Mediterranean offshore

AREAS OPENED TO EXPLORATION IN THE MEDITERRANEAN OFFSHORE

bitive. The issue was, in any case, settled in 1982 by the International Court of Justice in The Hague.

The Libyan offshore is almost entirely unexplored, mostly because of political problems. Dividing lines have yet to be defined between Libya and Malta, and there is a large zone of the seabed for which both countries have issued exploration permits. This area includes, among others, the Medina Bank, over which an Italian rig was drilling in the summer of 1981 under a permit issued by Malta when Libya sent a submarine to contest and stop the operations.

Apart from this, a number of companies have been dissatisfied with the economic conditions under which they operate in Libya, and in 1982 Exxon withdrew from the country. Added to this is the fact that there are large unexplored tracts onshore. These elements explain why the Libyan offshore has not yet been carefully explored, notwithstanding that from the geological point of view it is considered a very promising province - an extension of the same structures that are found onshore.

The promising part of the Libyan offshore is considered to include the Gulf of Sidra, while the portion extending north of Cyrenaica, where the continental shelf is very narrow, is not considered of interest.

The same applies to the Egyptian offshore west of the Nile Delta; although no wells have been drilled there, interest in this part of the Mediterranean offshore may change as the experts gain a better understanding of the geological formations below the Western Desert where important discoveries were made by Shell (Badr el-Din field). (9)

Considerable exploration is under way off the shores of the Nile Delta. The area of interest is considered to extend in front of the Sinai, but political and military conditions have prevented exploration activity there until very recently. Besides the al-Tina field (Agip and Total) that we already mentioned, a gas field was recently located by Mobil (el Temsah), which is added to the Abu Qir gas field that began producing in January 1979. Total also holds an exploration permit for a large area west of Alexandria called Maryout. (10) This is therefore an area in which further developments can be expected.

Israel engaged in some exploration activity in its own offshore, without success. No exploration has been carried out off the shores of Lebanon, Syria or Cyprus, because of political problems. Agip has been exploring in the Gulf of Iskenderun, without success.

As far as the Aegean is concerned, exploration has been discouraged by political considerations and by the mediocre geological prospects which promise meagre results, hardly worth the political controversy which exists.

Exploration is under way in the Ionian offshore of Greece, where the state oil company has made a gas discovery off Katakolon in the western Peloponnese; Agip recently took an exploration permit for the area around the island of Paxoi.

There is no record of exploration being carried out in the Albanian offshore. The country does not allow foreign companies to operate in its territory, but from the geological standpoint the prospects appear interesting.

Exploration in the Adriatic is, on the other hand, quite considerable. The delimitation of national rights over the seabed has been agreed upon between Italy and Yugoslavia. Exploration has been more intense on the Italian side because it was only recently that Yugoslavia modified its legislation and allowed for the participation of foreign companies in joint ventures. Numerous gas and oil fields have been found on the Italian side and exploration is now extending to the southern, deeper part of the Adriatic. The entire Adriatic region is geologically promising and can be expected to give rise to expanded production activity.

Some finds also occurred along the Ionian coast of Italy; the continental shelf is, however, very narrow there, as it is on the eastern side of Sicily. Exploration led to positive results to the south of Sicily, and a preliminary agreements exists between Italy and Malta on the delimitation of national rights. Exploration areas have been allocated by Malta, covering mostly areas north of the island. As was mentioned earlier, the whole area between Tunisia, Libya, Malta and Sicily is considered very promising, and exploration would be more intense if there were total agreement on delimitation of national rights. Unfortunately, this is far from being the case. At any rate, there is considerable activity throughout this area. In Italian waters south of Sicily exploration permits are held by Agip, Total and a consortium including Conoco, Exxon and Hispanoil. (11)

Although the continental shelf is very narrow in the

Tyrrhenian, some exploration was carried out along the Italian coast, yielding disappointing results. Drilling activity along the French Mediterranean coast, particularly in the Gulf of Lions, also brought no discoveries. Finally, exploration activity along the Spanish coast has been concentrated in the sector facing the Baleares, leading to the discoveries already mentioned.

In the spring of 1982 there were 28 offshore drilling rigs at work in the Mediterranean (out of a world total of some 470), ten of which were operating in Italian waters.

To conclude this overview of the hydrocarbon production situation and potential in the Mediterranean, it can be noted that, although exploration activity has been intensified in recent years, it could and perhaps will be increased even further, political conditions permitting. If it were not for political constraints, exploration would certainly intensify in those areas of the central and eastern Mediterranean where the seabed lies under less than 3,000 feet of water.

But a real turning point - in terms of both economic and political impact - will be reached only when it becomes technologically feasible to more actively explore the deep sea plains under more than 3,000 ft. of water. These are the largest part of the sea in terms of total surface, and are as yet practically untouched because of both technological and political problems. Seismic exploration is limited by the presence of a thick layer of salt, which acts as a screen and conceals formations underneath it. Actual drilling will therefore be necessary in order to gain a sufficient understanding of the structure of the deep Mediterranean basin. The data currently available give a contradictory picture of the potential: some elements are extremely encouraging while others point in an opposite direction. (12) The situation is considered, at any rate, sufficiently interesting to justify further exploration, and the fact that this is not taking place in any significant way is attributable to political obstacles - in view of the high costs and the degree of risk involved.

OIL TRANSPORTATION IN THE MEDITERRANEAN

Determinants of Mediterranean Oil Traffic

The total volume of oil which is shipped across the Mediterranean is influenced by three main variables: a) the level of production in the Mediterranean oil-exporting countries; b) the level of consumption in the Mediterranean oil-importing countries; c) the pattern of the transportation system connecting exporting countries and importing countries that lie outside the Mediterranean.

While it is clear that oil that is exported (imported) from (to) a Mediterranean country must necessarily transit in the Mediterranean even if it is directed to (comes from) a country outside the Mediterranean, the same is not true for the case contemplated under c).

Although the flows generated by cases a) and b) should certainly not be neglected, I do not intend to propose yet another projection of future oil demand and supply. A few data on the production level in the Mediterranean exporting countries will suffice (Table 1.4). It should be noted that in the case of Libya and Algeria the maximum rated production capacity is greater than actual production (in the case of Libya very much so). However, it is unlikely that production will approach maximum capacity in these countries during the current decade. On the other hand, production in Tunisia and Egypt will tend to be much closer to maximum capacity, and in the case of the latter is expected to grow rather rapidly, reaching a level of one million b/d by 1985. (13)

The Redirection of Oil Logistics and the Mediterranean

What is affecting most remarkably the importance of the Mediterranean in the broader picture of oil flows is, however, a number of changes being introduced or planned for the logistics of oil. In essence, these changes are designed to reduce the present and necessary reliance on transportation across the Persian Gulf, the Strait of Hormuz and around the African continent - a route which is seen as being increasingly vulnerable to a wide array of possible threats (ranging from non-governmental use of force to interference from regional or outside governments).

The tendency is in the direction of recreating the balance which existed until 1967. The Mediterranean was historically seen as the natural outlet for Middle East oil, witness the very considerable diplomatic activity surrounding the establishment of the IPC pipeline, connecting the northern Iraqi fields to the ports of Banias, Tripoli and Haifa. The exact routing of the line, as well as its multiple outlets, was the result of complex negotiations between the U.K. and France in 1919; (14) at the time a pipeline connecting the Persian fields to the Mediterranean was also being considered. (15)

Table 1.4: Production of crude oil in selected Mediterranean countries (million tons/year)

	1960	1965	1970	1975	1980
Algeria	8.5	26.0	47.3	45.1	44.9
Egypt	3.3	6.5	20.9	11.7	30.0
Libya	-	58.8	159.2	72.4	85.6
Tunisia	-	-	4.2	4.6	5.2
TOTAL	11.8	91.3	231.6	133.8	165.7

Source: ENI, Energia ed Idrocarburi, 1981, p. 62.

The development of oil in Saudi Arabia led to the establishment of another pipeline to the Mediterranean (the Tapline) connecting the fields around Dhahran to the harbours of Sidon and Haifa.

The creation of Israel led to a considerable disruption in this system. Not only was the flow of oil to Haifa interrupted, but the reality of continuing war on the eastern Mediterranean coast acted to discourage the expansion of this overland transportation system to the Mediterranean. The Tapline was repeatedly interrupted by terrorist attacks.

A further discouragement came from recurring conflict over the fees that were to be paid to the countries whose territories the pipeline crossed. The best known episode is the one which saw Syria opposed to the IPC in 1966-67, leading to a stoppage of the oil flow. (16) Tensions resurfaced in the 1970s after Iraq nationalised the IPC, proving that in itself nationalisation was no solution to the problem.

As a consequence, a larger portion of the oil came to be loaded on ships directly in the Gulf; but the prevalent route for the tankers was still through the Suez Canal and across the Mediterranan.

However, in 1967, Israeli forces occupied the Sinai and the Canal was closed. This accelerated a tendency to increase the size of new tankers and rely on the route around the Cape. It must be stressed at this point that Very Large and Ultra Large Crude Carriers (VLCCs and ULCCs) raise many more problems than just transit through the Suez. Besides being more accident prone and a greater hazard to the environment than the medium-size tankers are, their economic advantage was substantially eroded during the 1970s because of the increase in the price of oil itself for they are "bundker guzzlers". Furthermore, many major harbours around the world are still not equipped to accommodate these ships; it was only in 1982 that the so-called LOOP became operational in Louisiana, the first American harbour capable of receiving ULCCs. In other words, it is likely that, had the Suez Canal not been closed by the events in 1967, the average size composition of the tanker fleet would not have increased as dramatically as it did. In the early 1970s many of those tankers were lying idle, expectations within the industry were gloomy, and considerable scrapping was under way. (17)

Developments in Iraq in the 1970s

After the 1967 war, a further development that affected the logistics of oil was Iraq's attempt to reduce its dependency on the Syrian pipeline. This was accomplished by creating two further pipelines.

A first pipeline runs from north to south, and is sometimes called "strategic". Its purpose is to establish a line of communication between the northern Iraqi fields around Kirkuk and the southern ones around Basrah. Hitherto, Kirkuk oil could only be shipped across Syria, and Basrah oil could only be loaded on the Gulf at the mouth of the Shatt-al-Arab. Both outlets were (and are) unreliable, in the case of Syria because of the recurring conflicts over the level of transit fees (which occur in the context of a generally conflictual relationship at the economic as well as at the political level), and in the case of Iran because of the

17

border dispute over the mouth of the Shatt (itself occurring in the context of tense relations because of the Kurdish problem, the Shia opposition, etc.). It was only in 1975 that the dispute over the Shatt was "solved" by an agreement between the Shah and Saddam Hussein, only to flare up once again at the end of 1980, as Iraqi troops crossed the river and marched into Iran. (18)

The outbreak of the Iran-Iraq war brought Iraqi shipments from the Gulf to a complete halt and a determined effort was made to utilize the pipeline to Banias and Tripoli. Theoretically, this pipeline has a maximum capacity of 1.4 million b/d, but it had been allowed to deteriorate considerably over the years (the Tripoli terminal, for example, had been inactive since 1975). Thus the operational capacity at the beginning of 1982 was considered to be around 800,000 b/d. The Iraqi efforts have not been entirely successful because of repeated damage done both by Iranian air attacks and by terrorist bombings. However, the line was being used to export some 400,000 b/d immediately before 10 April 1982, when Syria unilaterally decided to close it, while at the same time starting to import Iranian oil. (19)

The "strategic" pipeline is characterised by the fact that it is capable of operating in both directions, thus making it possible to export Kirkuk oil from the Gulf and Basrah oil from the Mediterranean. If it operates southward it has a maximum throughput of one million b/d, while if it operates northward maximum capacity is 800,000 b/d; the difference is due to the fact that in the second case it operates against gravity. Since its opening, it has operated mostly in the southward direction, until the beginning of hostilities with Iran in 1980.

A second pipeline was built connecting the Kirkuk fields to an outlet at Ceyhan (Turkey), without crossing Syria. It has a maximum throughput of 700,000 b/d, has been operational since 1978, and has played a vital role for Iraq since the beginning of hostilities with Iran. Its maximum throughput could easily be increased to 900,000 b/d. (20)

Although both of these pipelines were mainly intended to reduce Iraqi dependence on Syria, they also happened to allow a reorientation of Iraqi oil flows from the Gulf to the Mediterranean, which actually took place dramatically after September 1980.

The Saudi East-West Pipeline

Out of concern for political and military conditions in the Gulf and the vulnerability of the oil flows at Hormuz, Saudi Arabia has also installed a "strategic" pipeline running from east to west and connecting the Saudi fields to the Red Sea port of Yanbu. This pipeline became operational in mid-1981 and has a maximum throughput of 1.85 million b/d.

It was initially expected that this pipeline would allow a further lowering of the transportation cost for Saudi crude, particularly for destinations in the Mediterranean. However, the Saudi authorities set the transportation fee at a relatively high level (60 cents a barrel) to amortize the investment speedily, and buyers of Saudi crude had to be pressured into using the pipeline. The Aramco partners asked Saudi Arabia to average out the 60 cents and sell at the same price both at Yanbu and Ras Tanura, allowing the companies to choose freely where to load the oil from. (21) This request was not granted, but at the beginning of 1982 Petromin started "waiving" the requirement to lift at Yanbu and utilization of the line dropped instead of progressively increasing to maximum capacity.

This situation should, however, be understood in the context of a generally weak oil market and declining Saudi output. It is very likely that the Saudi interest in the East-West pipeline was not in the least reduced because of the less than enthusiastic reception that it met initially.

It is therefore appropriate to record that while, as we said, the present maximum capacity of the line is 1.85 million b/d, consideration has already been given to the possibility of looping the pipeline to raise its capacity to 2.45 million b/d. Further improvements could bring the capacity of the line to 3.7-4 million b/d, which would certainly provide substantial flexibility. (22)

There has also been some speculation that Saudi Arabia might be considering the creation of a massive strategic oil reserve near Yanbu (1.5 billion barrels). This was prompted by the fact that a company was asked to study underground caverns in the proximity of Yanbu, that could be used for storage. (23) Such speculation was promptly denied by Saudi Oil Minister Yamani; however, the creation of such a reserve would appear to be a logical step, and, if it were being considered, or actually accumulated, one would expect the Saudi authorities to deny its existence and to maintain it secret.

The Mediterranean and The Energy Picture

The Sumed and the Reopening of the Suez Canal

The partial withdrawal of Israeli forces from the Sinai allowed Egypt to reopen the Suez Canal on 5 June 1975. While the Canal was closed, Egypt had endeavoured to provide an alternative to oil transportation around the Cape through the laying of a pipeline connecting the Red Sea to the Mediterranean - the Sumed. The Sumed has a maximum throughput of 1.6 million b/d, and has been utilized very close to its maximum capacity.

Since it reopened, the Suez Canal has been both deepened and enlarged. A first phase of this process was concluded at the end of 1980 and resulted in a considerable increase in traffic which went to 225 million tons during the first eight months of 1981 from 187 million tons a year earlier. With the first phase the draught in the Canal was increased from 38 to 53 feet which is sufficient to allow for the transit of fully laden tankers up to 150,000 tons and of VLCCs up to 375,000 tons in ballast. The increased traffic consisted initially mainly in southbound tankers transiting in ballast, because with depressed tanker rates the cheapest combination is that of loading oil in large tankers that will follow the Cape route to the European market and return south via Suez. Table 1.5 shows some estimates of the relative cost of various options as of April 1981.

However, the relative importance of the two currents of traffic has changed since the opening of Saudi Arabia's pipeline to Yanbu. In December 1981, northbound laden tankers jumped to 5.7 million tons, from 1.3 million tons in the previous month. (24)

For the year 1981 the total number of tanker transits was 21,577, equal to a daily average of almost 60.

The importance of the Suez Canal to oil traffic may further increase in the future if a further deepening is carried out. In this case the draught would increase to 67 feet and allow passage of fully laden tankers of up to 260,000 tons. (25) There are, however, doubts as to the economic return of a further deepening, because the spectrum of potential traffic increase is rather narrow (fully laden tankers above 150,000 tons and below 260,000 tons), and because it is expected that the average tanker size will decrease in the future. A further deepening of the Canal must be weighted against a possible enlargement of the Sumed, also considering that tankers in the 150-260,000-ton

20

Table 1.5.: Estimated spot tanker freight costs for Arabian Light to representative destinations from either Yanbu or Ras Tanura, various transportation alternatives. Period of reference: April 1981 (in dollars per barrel)

6

	Rotterdam		Sicily		Philadelphia		Curaçao	
	Yanbu	R.T.	Yanbu	R.T.	Yanbu	R.T.	Yanbu	R.T.
Direct Voyage, ULCC, Cape round trip	0.90	0.94	0.90	0.94	1.35	1.39	0.82	0.86
Direct Voyage, VLCC, Cape and Suez return	0.85	1.00	0.75	0.90	1.39	1.54	1.00	1.04
Suez, round trip, 150,000 tonner	0.86	1.22	0.56	0.90	1.61	2.23	1.09	1.46
Sumed, ULCC	0.65	0.89	0.49	0.73	1.58	1.82	0.85	1.09

Source: Petroleum Intelligence Weekly, 27 April 1981, p. 2.

class can already transit provided they are not fully laden (in other words, they can unload part of their cargo at one end of the Sumed, transit the Canal partly laden and reload in the Mediterranean).

Plans to enlarge the maximum throughput of the Sumed are more solid. Since the opening of the Yanbu pipeline, the Sumed has been operating above or close to 100% capacity and the average utilisation for the year 1981 was 96%. (25)

It should be remembered that the Yanbu pipeline is expected to reach maximum capacity only in the second half of 1982.

The capacity of the Sumed could certainly be raised to 1.9 million b/d and possibly further to 2.3 million b/d. Approximately 65% of the crude transported through the Sumed is Arabian Light.

In the longer run, the relative importance of Suez vs. the Cape route should be increasing also in connection with the tendency of the oil-exporting countries to process a larger share of their crude oil at home and export products rather than crude oil. Products are normally shipped in smaller vessels than crude is. Furthermore, the tendency of the oil-producing countries to diversify their oil logistics out of the Gulf is not exhausted, and there are plans for more pipelines that would bring an increase in Suez traffic. To these we now turn.

New Strategic Pipelines Under Consideration

Persistent conditions of tension in the Gulf are encouraging the oil-exporting countries in the region to further diversify their oil logistics. We already mentioned that the capacity of the Saudi East-West pipeline could be increased, and the same applies to the Iraqi-Turkish line. Indeed, a protocol between Iraq and Turkey was signed in Ankara in December 1980 carrying an agreement to increase the capacity of the line to 1 million b/d. (26)

However, there are also substantial plans for altogether new pipelines, some of which would most likely lead to an increase in oil traffic through the Mediterranean. Specifically, Iraq is considering two new lines, one across Kuwait to a loading terminal in the Gulf, the second across Saudi Arabia to the Red Sea. (27)

The Iraqi plans are probably very serious, particularly since Syria unilaterally closed down the pipeline across its territory in April 1982, dealing a final blow to the year-

long effort to recover export volume that it had lost
because of the war with Iran.

A pipeline across Kuwait would specifically reduce
Iraqi dependence on the Shatt-al-Arab as its only outlet to
the Gulf. It would not, however, amount to a significant
diversification from the Gulf, and it would have no impact
on traffic through the Mediterranan.

On the other hand, the proposed line across Saudi
Arabia to a point slightly north of Yanbu would be a
significant factor for Mediterranean traffic. The Iraqi Oil
Minister Abdul Karim stated in March 1982 that work on this
line could start by the end of 1982 and would take no more
than two years to complete. Its total capacity would be
somewhere between 1 and 1.6 million b/d. (28)

Future Evolution of Oil Traffic Across the Mediterranean

On the basis of the developments listed above, the
importance of the Mediterranean for international oil traffic appears destined to increase greatly, both in normal
conditions and in an emergency.

The latter case is more easily discussed. On the
expectation of continuing tension in and around the Gulf, we
may assume that pipelines presently being considered will actually become operational. In the event of a total stoppage
of Gulf shipments, this would leave a maximum capacity of
2.4 million b/d directly to the Mediterranean, and of 5
million b/d to the Red Sea. We may expect that most of the
oil available on the Red Sea would enter the Mediterranean
through either the Sumed or the Suez Canal.

Assuming that the Sumed is fully operational with a
maximum capacity of 2.3 million b/d, the Suez Canal would
bear traffic of around 2.7 million b/d. This figure is very
close to the projection of Suez traffic in 1985 that was put
forward by Internaft at the end of 1980. (29)

If we add the potential exports of Mediterranean producers, which we may estimate at around 3 million b/d, we
reach a total of more than 10 million b/d that would be
available in the Mediterranean if the Gulf were completely
closed to naval traffic in an emergency (see Table 1.6).
This would cover some 50% of US and West European imports in
1985.

Outside an emergency, the volume of traffic would be
influenced by a variety of elements, most of which are
impossible to predict. However, if pipelines were laid they

would tend to be used in non-emergency times as well, causing an increase in Mediterranean traffic anyway. A figure of 8 million b/d for Mediterranean oil shipments in 1985 is therefore certainly plausible.

Table 1.6.: Estimate of Mediterranean crude oil shipments in 1985

	in million b/d
Sumed	2.3
Suez Canal	2.7
Iraq-Turkey	0.9
Iraq-Banias Iraq-Tripoli	1.4
Libya*	1.0
Algeria*	0.8
Tunisia	0.3
Egypt	1.0
TOTAL	10.4

* Maximum sustainable output is estimated today at 2.1 million b/d for Libya and 1.2 million b/d for Algeria.

THE MEDITERRANEAN AND THE DEVELOPMENT OF NATURAL GAS RE-SOURCES

Some Peculiarities of Natural Gas

If the Mediterranean is likely to become increasingly important in the geography of the international oil industry, its significance for the development of natural gas may turn out to be even greater, all the more so because of the predominantly regional role of natural gas as an energy source.

The main obstacle to exploitation of natural gas is transportation. Delivery to the consumer necessarily requires the creation of a pipeline grid reaching each individual customer. This highly capital-intensive distribution system can be fed by local natural gas, if it is available;

24

if not, natural gas has to be transported from the source to the inlet of the distribution grid. This can be done either in gaseous form - through a pipeline - or in the form of liquefied natural gas (LNG). In the latter case, which is relevant only for maritime transportation, the gas is liquefied in a liquefaction plant, is transported in special LNG tankers, and is then regasified at the inlet of the distribution system.

The difficulty of transportation is much increased whenever gas must be transported across the sea. There are economic as well as technological limits to the laying of pipelines on the seabed at great depths, while the LNG route is costly both in terms of initial investment and in terms of direct costs. This is the fundamental reason why the share of total energy needs which is met by natural gas varies greatly, depending on the local availability of gas. Thus, in the USA natural gas accounts for around 25% of primary energy, while in Western Europe for only some 14%.

The difficulty of transportation is the main reason why, in the past, international oil companies did not bother to develop the natural gas reserves found in the oil-exporting countries. The sheer waste is now beginning to disappear (only too slowly). However, the issue of full development of gas resources has not yet been systematically tackled.

Oil-exporting countries have often expressed a preference for the local utilisation of natural gas resources: oil may be exported, but gas is one asset that should be kept for local use. There is controversy over the netback value that local utilisation of gas would bring relative to the price at which it may be exported. However, it would appear that reserves available in many countries are so large relative to all realistic estimates of potential local demand that even if there were an absolute preference for local transformation, significant quantities would still be available for export. A total ban on exports would prolong the life of known reserves to an irrationally distant horizon.

If natural gas is offered for export in large quantities, it will become a major factor in Mediterranean trade. Although the shipment of gas in LNG form over great distances is not to be ruled out, most of the gas trade will probably remain "regional". The reason for this expectation is that sufficient discoveries are being made in the North

25

American continent to make it potentially self-sufficient –
and the high cost of transportation will ensure that the
closer sources are tapped first. For the same reason, Japan
will tend to rely on sources in southeast Asia, although it
is to be expected that it will also import some LNG from the
Gulf. Most of the gas available for export from the countries around the Gulf, from North Africa and Nigeria will
therefore be shipped to European outlets. With the exception
of potential LNG exports from Nigeria, which could be
transported without ever touching the Mediterranean, all
other flows would somehow cross the Mediterranean, either in
gaseous or in LNG form.

There is considerable uncertainty as to the future
size of this trade. Although available resources are very
large (see Table 1.7; in considering these data it should
not be forgotten that in most countries gas has not been
systematically explored for), the growth of demand in the
industrial countries may be considerably slowed down by the
insistence of the producing countries on the equalization of
the price of gas and crude oil. Therefore, we will not
discuss forecasts of the importance of Mediterranean gas
trade at some specified future date, but will describe the
existing export and import infrastructures in the Mediterranean as well as various projects that are being considered and may, sooner or later, be implemented.

Current Infrastructure for Gas Trade in the Mediterranean

The Mediterranean country which has made the greatest
effort to promote gas exports is Algeria. The Algerian infrastructure for gas exports consists of plants for gas
liquefaction and the pipeline connecting the Hassi R'Mel
field to Italy, across Tunisia and the Sicilian Channel.

Table 1.8 shows the global situation of existing LNG
plants in operation in 1980. The leading role played by
Algeria is immediately evident.

The Algerian experience is thus extremely interesting,
but suggests that liquefaction poses numerous problems. (30)
In fact, in 1981, the Algerian government revised its commitment to the continued development of this industry, indefinitely postponing plans for a third liquefaction complex
in Arzew.

Table 1.7: Natural gas reserves around the Mediterranean (billion cubic metres)

Country	1970	1979	1980
Algeria	3,000	3,738	3,724
Libya	850	680	674
Nigeria	170	1,172	1,161
Other African countries	400	367	345
Abu Dhabi	-	538	566
Saudi Arabia	1,500	2,711	3,183
Iraq	600	779	777
Iran	6,000	13,877	13,735
Kuwait	1,000	949	940
Qatar	-	1,694	1,700
Other M.E. countries	750	413	407
TOTAL	14,270	26,923	27,212

Source: Eni, Energy and Hydrocarbons, 1981, p. 56-58.

In the Mediterranean context, the export of gas via pipeline may have a cost advantage over LNG and the higher the price of gas the more relevant is the advantage. For this reason much importance was attached to the Transmed pipeline as a precedent for the more efficient export of gas across the sea. Unfortunately, Algeria's insistence on a price system that would deliver gas to the Italian network at the same cost as the LNG alternative would have led to is considerably reducing the long-term implications of this new venture.

The Transmed pipeline represents a technological break-through not so much because of its total length (about 2,500 km), but because it has been laid at depths of up to 600 metres and traverses a fairly rugged seabed (Fig. 1.4 and 1.5). The pipeline has a maximum capacity of 12.4 billion cubic metres per year.

Most of the Algerian LNG is delivered outside the Mediterranean. France is the only country on the Mediterranean

Figure 1.4 Transmediterranean pipeline

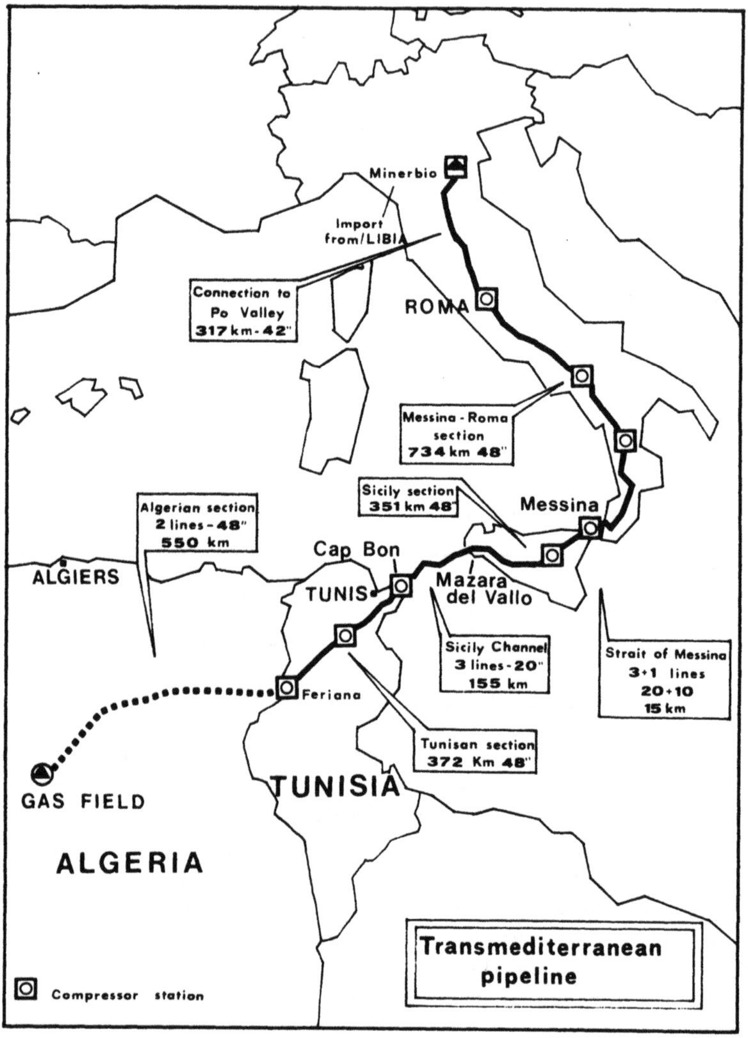
Minerbio
Import from/LIBIA
Connection to Po Valley 317 km - 42"
ROMA
Messina - Roma section 734 km 48"
Sicily section 351 km 48"
Messina
Algerian section 2 lines - 48" 550 km
Cap Bon
ALGIERS
TUNIS
Mazara del Vallo
Sicily Channel 3 lines - 20" 155 km
Strait of Messina 3+1 lines 20+10 15 km
Feriana
Tunisan section 372 Km 48"
GAS FIELD
TUNISIA
ALGERIA
Transmediterranean pipeline
Compressor station

Figure 1.5   Strait of Messina and Sicily Channel crossings

STRAIT OF MESSINA CROSSING

(depth in metres)

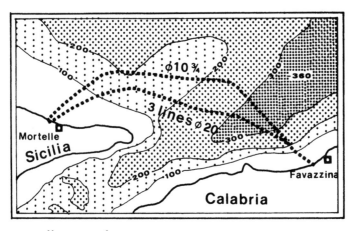

sea-line route

SICILY CHANNEL CROSSING

(depth in metres)

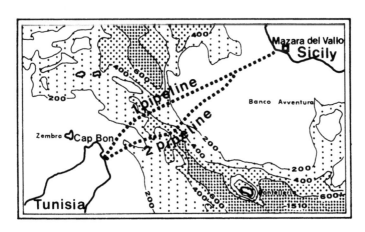

sea-line route

The Mediterranean and The Energy Picture

Table 1.8: Operational LNG projects at the end of 1980

Exporting country	Market area	Start-up year	Contract volume (billion cubic metres)
Abu Dhabi	Japan	1977	4,0
Alaska	Japan	1969	1,4
Algeria	United Kingdom	1964	1,0
Algeria	France	1965&72	4,0
Algeria	USA	1976&77	11,2
Algeria	Spain	1976	4,5
Total Algeria			20,7
Brunei	Japan	1972	7,5
Libya	Italy	1972	2,7
Libya	Spain	1972	1,2
Total Libya			3,9
Indonesia	Japan	1977	11,0
TOTAL			48,5

Source: <u>Energy and Hydrocarbons 1981</u>, ENI, p. 87

coast which imports LNG. The terminal is at Fos sur Mer.

Libya is currently the only other Mediterranean country besides Algeria which possesses infrastructure for the
export of gas, consisting in a relatively small LNG plant at
Brega, formerly owned by Exxon, from which gas is normally
exported to a terminal in Italy (La Spezia) and one in Spain.

Outside the Mediterranean, there are plants for the
production of liquefied petroleum gas (LPG) in a number of
Gulf countries, and LNG in Abu Dhabi, but their output   is
at present mostly absorbed by the Japanese market and does
not enter into Mediterranean trade.

Possible Developments of Gas Trade in the Mediterranean

The most dramatic potential developments in the Mediterranean gas trade are linked to the exploitation of resources in Iran and in the Arab Gulf countries. From the
technological point of view, three alternatives exist: 1)
liquefaction close to the source (i.e. on the Gulf) and shipment in LNG tankers through Suez to a Mediterranean or north
European terminal; 2) transportation by pipeline to a point
on the Mediterranean coast, then liquefaction and transport
in LNG form to a Mediterranean or north European terminal;
3) transportation entirely by pipeline.

The alternatives are depicted in Fig. 1.6, while Table
1.9 summarizes a cost comparison for each alternative under
a uniform set of assumptions. One should notice that the
assumption on quantity at the inlet of the system is important, because the investment costs are not proportional to
the quantity transported; the pipeline alternative becomes
more competitive the larger the quantity to be transported.
Also, the time distribution of the flow is important, the
pipeline being better off if the flow   is   constant.
Finally, as the table shows, the value of gas as fuel is
important, as the pipeline absorbs far less of it.·

The results of the cost comparison thus depend on the
assumptions; yet it is important to note that for a reasonable set of assumptions - such as those used by Bonfiglioli
and Cima - the pipeline alternative appears preferable even
for gas from the Gulf. Yet this certainly does not authorise
the conclusion that gas will be exported to Europe only via
pipeline.

The most likely outcome is that a system for transporting gas from the Gulf to Europe will evolve gradually and
in a complex way. No country will want to rely on just one

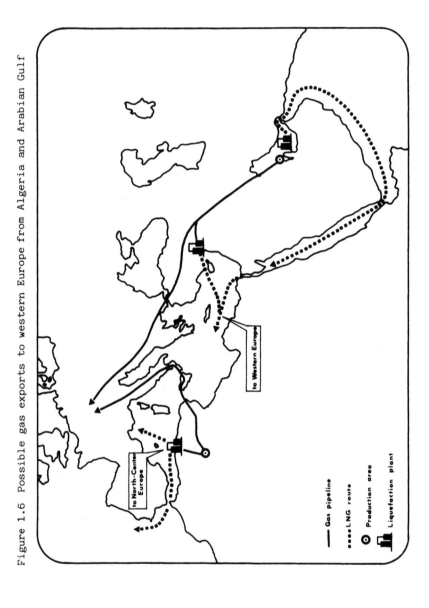

Figure 1.6 Possible gas exports to western Europe from Algeria and Arabian Gulf

source and one mode of transportation. Liquefaction plants will be opened because they can be decided upon unilaterally by each state, and because they allow a shift in the final destination of LNG. A pipeline system collecting gas from various sources around the Gulf is a difficult political proposition. It is more likely that Iran or Iraq, or both, will take the initiative to build a pipeline across Turkey to liquefaction plants on the Mediterranean (31) - and economic reasons would not be the only consideration. Thence, the system may expand, with other Arab countries possibly connecting into the Iraqi line, and European countries promoting a feeder from a gathering point in Turkey. Although this appears to be a realistic scenario, it may very well be that political difficulties and an unrealistic price stance on the part of the gas producers will lead to a situation whereby gas exports to Europe simply fail to happen, or never grow beyond a low level. In the latter case, LNG would remain the only option.

Yet, even if the Middle East is the largest potential source of natural gas for Mediterranean trade and its future is highly uncertain, there are other important gas reserves elsewhere around the Mediterranean, in particular in Egypt and Tunisia.

As a policy measure, Egypt has decided to retain natural gas for domestic use until a minimum of 345 billion st.cu.mt. (12 trillion st.cu.ft.) of reserves are located, which will consititute the so-called "National Reserves". Once this reserve has been secured, a gas export project will be allowed, with the possibility of pooling the gas reserves of the different operators. (32) The export of gas will most likely take place in LNG form; considering the present reserves and the time distribution of the likely discovery of new reserves, we can visualize the possibility of an LNG plant producing 3 billion cu.m. initially in 1990, whose capacity would be increased to 9 billion cu.m. in 1991 and stabilised at this level. (33)

Another possible infrastructural development for gas trade in the Mediterranean is directly or indirectly linked to Algeria. Before the Transmed pipeline was completed and the row over the price of gas began between Italy and Algeria, the following developments had been under consideration: a) an increase in the capacity of the Transmed to 18 billion cu.m./year or more (34); b) the laying of a pipeline across the Otranto Channel, connecting Greece to the

Table 1.9: Cost comparison of alternative modes of transporting natural gas

	Algeria-Europe		Arabian Gulf-Europe		
	Gas pipeline	LNG	Gas pipeline	LNG	Gas pipeline LNG
	Hass i R' Mel-Tunisia-Italy-northcentral Europe	—	Gulf Area-Iraq-Turkey-Greece-Yugoslavia-northcentral Europe	Via Suez	Liquefaction at Mediterranean coast
Length of route (Km)	3,500	3,350	5,200	11,300	7,000
- of which land gas pipelines	3,320	550	5,200	–	2,300
sea gas pipelines	180	–	–	–	–
Quantity at the inlet of the system (10^9 m³/a)	18	18	18	18	18
Quantity delivered to market (10^9 m³/a)	16.8	15.1	16.3	14.8	14.4
Investment at April 1980 costs without interest during construction (10^9 $)	4.5	4.5	5.9	6.5	7.1

Transportation unit cost
at April 1980 value ($/MMBtu)

a) fuel gas value 2 $/MMBtu	1.5	2.1	2.0	2.9	3.1
b) fuel gas value 3 $/MMBtu	1.6	2.3	2.1	3.1	3.3
c) fuel gas value 4 $/MMBtu	1.7	2.5	2.2	3.3	3.5

Source: Bonfiglioli e Cima, op. cit., page 38

Transmed; c) the laying of a second Transmediterranean pipeline connecting Algeria to Spain (from Oued Mellah to
Almeria). (35)

Offshore Gas Reserves in the Mediterranean

We have so far discussed the relevance of the Mediterranean for regional gas trade. The picture is, however,
incomplete without consideration of offshore gas reserves in
the Mediterranean. The general conditions of offshore hydrocarbon exploration and production in the Mediterranean with
reference to oil have already been described, and we need
add only a few specific remarks on gas.

A first important point is that prospects for gas in
the Mediterranean are generally better than for oil, as is
borne out by the fact that many more gas fields have been
found than oil fields. Numerous gas fields have been located
in the northern Adriatic on the Italian side. Large gas
reserves have also been discovered off Tunisia and Egypt. In
the former, the Miskar field is estimated by Etap to contain
30 billion cubic metres of gas. In Egypt the largest offshore gas find is at Abu Qir, which began normal production
in January 1979 at a rate of 100 million cu. ft/day. (36)
Plans for a doubling of this capacity and further improvements have been considered for some time (37) and financial
assistance to this end was recently offered by the World
Bank.

One important aspect of offshore gas in the Mediterranean is the exploitation of minor discoveries. At present,
an offshore gas field can be exploited only if it is
connected to the coast by means of a pipeline; and in view
of the generally difficult conditions in the Mediterranean
many gas discoveries may simply not be worth the investment.
If a small oil deposit is found offshore, it is always possible to exploit it with minimum investment through the so-
called "early production" method. This means that the oil is
pumped directly from the field into a tanker which is
temporarily anchored nearby and serves as a floating reser-
voir from which other tankers may load. However, this is not
possible with gas.

In order to exploit these fields it has therefore been
suggested that gas may be utilized in situ with floating
plants that would connect directly to the field. These would
most probably be power plants, which would be connected to
the national grid by means of underwater cables.

Although this is at present a largely theoretical hypothesis, it is important that it be mentioned for it would entail a considerable increase in the economic interests at stake in and around the seas.

Conclusion

The various developments that have been analysed so far point to a growing importance of the Mediterranean in the global energy picture. However, while on some aspects it is possible to make forecasts with relative confidence, on others there is still considerable uncertainty.

Two processes under way stand out clearly: a) exploration and development of oil resources in water depths of less than 3,000 ft.; b) transportation of oil to and across the Mediterranean in growing quantities.

In political terms, the first process may be a source of conflict, and indeed has been. However, it would appear that the interests at stake, while significant, are still relatively minor, and the chances that a peaceful compromise can be found are considerable. In military terms, offshore oil installations would be mostly along the coasts, and would not interfere with - or be harmed by - the presence of military forces at sea. A possible exception to this conclusion relates to the Sicilian Channel, which is at the same time an important choke point and an important area for oil activities.

The second process underlines the need to maintain the role of the Mediterranean as an open sea for international navigation. It is a process that would not imply any significant change in the present political and strategic status quo. It would, rather, tend to reduce the importance attached to the Gulf and the Indian Ocean, most notably to the extent that there is a trade-off between naval superiority in these areas and in the Mediterranean.

Two further processes are possible: a) the development of natural gas trade; b) the extension of oil exploration and production to areas covered by more than 3,000 ft. of water.

The development of natural gas trade would increase the economic and strategic importance of the sea very considerably. Because of its properties, gas is likely to continue as an essential component of the energy balance even in the relatively distant future; and Mediterranean gas trade is the primary alternative to Western Europe's growing direct dependence on gas supplies from the Soviet Union. All elements of the LNG chain are considerably more vulnerable to military action than corresponding installations for oil trade. It is also more difficult to maintain gas stocks that would compensate a temporary reduction in shipments - except in the form of domestic reserves which are deliberately not produced. Finally a multiplication of underwater gas pipelines would create a strong incentive for the countries concerned to establish a strong military control over specific sections of the sea, possibly limiting the presence and passage of forces of other countries.

The extension of oil exploration and production to the deep basins would alter the political and military environment very considerably. Conflict over the delimitation of underwater boundaries may become very serious, in view of the difficulty of reaching "fair" decisions in the Mediterranean context.

Furthermore, the operations would involve very sophisticated equipment and technology, which would almost certainly impinge upon the strategic balance in the sea. Not only would civilian installations have to be protected; they may themselves acquire a military significance (notably in connection with detection of submarine traffic).

Hence the importance of developing a cooperative approach to the development of resources in the deep basins. The fact that a lot of technological development and geological research is needed before these resources will acquire an immediate economic significance should enhance the chances of reaching an agreement among a large number of Mediterranan countries to approach the problem jointly. The need to precisely delineate underwater boundaries might be overcome in this context, and conditions might be more favourable to the disentanglement of civilian and military objectives and instruments.

NOTES

1. Total Information, No. 78, page 33.
2. Offshore, 5 June 1981.
3. Petroleum Intelligence Weekly, 9 April 1981, p. 12; Business Week, 15 March 1982, p. 36.
    4. Petroleum Intelligence Weekly, 9 April 1981, p. 10.
    5. Annuaire Européen du Pétrole 1981/82, Hamburg.
    6. Petroleum Intelligence Weekly, 11 January 1982, p. 5.
7. Ibid., 22 March 1982, p. 7.
8. Ibid., 15 June 1981, p. 11; Ecos, No. 100-101, 1981, p. 15.
    9. Petroleum Intelligence Weekly, 19 April 1982, p. 9.
    10. Total Now, March 1982.
    11. Esso Informazioni Economiche, No. 1, 1981.
    12. These are briefly spelled out by Bernard Biju-Duval in "Deep basins: promising or unpromising" (Total Information, No. 81, 1980, p. 18-20).
    13. Petroleum Intelligence Weekly, 19 April 1982.
    14. Marian Kent, Oil and Empire, British Policy and Mesopotamian Oil 1900-10-20, Macmillan, London, pp. 141-150.
    15. Ibid., p. 143.
    16. For a detailed discussion of this see G.W. Stocking, Middle East Oil, Vanderbild University Press, Kingsport, 1970, chapters 12-13, pp. 270-299.
    17. Intertanko expected that a supply-demand balance for VLCCs could be reached by 1984-85 only if one-fourth of the global fleet were scrapped. In 1980, 26 VLCCs were sold for scarp and another 23 as of mid-July 1981 (Petroleum Intelligence Weekly, 7 September 1981, p. 10). While independent owners were selling large tankers for scrap, oil companies were starting to question the need to own a tanker fleet at all. In the first quarter of 1982, oil companies sold nine VLCCs for scrap, while the total for 1981 was eleven, "Oil companies lose money even on modern ships. And many would require further substantial investment to cut their fuel consumption and to comply with new international safety and anti-pollution requirements". (Petroleum Intelligence Weekly, 19 April 1982, p. 2).
    18. These historical details must be recalled lest the reader should think that once the Iraq-Iran conflict is resolved Iraq will again be happy with shipping its oil from the Gulf.

19. Middle East Economic Survey, 19 April 1982, p. 4.

20. Middle East Economic Survey, 3 May 1982.

21. Petroleum Intelligence Weekly, 19 October 1981, p. 3.

22. Ibid., 27 April 1981; Middle East Economic Survey, 9 November 1981, p. 6.

23. Petroleum Intelligence Weekly, 20 October 1981, p. 1.

24. Petroleum Intelligence Weekly, 8 February 1982, p. 9.

25. Ibid, 16 November 1981, p. 10.

26. Middle East Economic Survey, 5 January 1981, p. 6.

27. Middle East Economic Survey, 10 August 1981, pp. 1-2.

28. Petroleum Intelligence Weekly, 22 March 1982.

29. World Oil and Tanker Outlook to 1986, Internaft Ltd., London, 1980.

30. For a detailed discussion of this experience see: Ali Belhadj, "The Liquefaction of Natural Gas in Algeria: Development, Technologies, Experience", paper submitted to Algiers symposium, pp. 329-352.

31. Indeed, in December 1980 Iraq and Turkey signed a protocol in Ankara agreeing, among other things, on "the construction of a natural gas pipeline to carry Iraqi natural gas to Turkey", Middle East Economic Survey, 5 January 1981, p. 6).

32. M.K. El Ayouty, "Exploration and Gas Discoveries in Egypt"; G. Rustici, "The Policy of the Egyptian Government for Promoting Gas Exploration and Exploitation", papers submitted to the International Seminar on Natural Gas and Economic Development, organised by EGPC and IEOC in Cairo, 26-27 February 1982.

33. M. Colitti, "Natural Gas and Economic Development", ECPC-IEOC Seminar.

34. Middle East Economic Survey, 21 January 1980, p. 5; 12 May 1980, p. 6.

35. Ibid, 12 May 1980, p. 6.

36. Ibid, 29 January 1979.

37. Ibid, 1 February 1980.

Chapter Two

MEDITERRRANEAN NON-ENERGY RESOURCES: SCOPE FOR COOPERATION
AND DANGERS OF CONFLICT

Gerald H. Blake

INTRODUCTION

For the purpose of this paper the Mediterranean Sea is
taken to include the Aegean and Adriatic Seas, but not the
Black Sea (which the United Nations Food and Agriculture
Organization combines with the Mediterranean for fishing
statistics). Thus defined, the Mediterranean occupies
2,511,000 sq.km. or 0.7 per cent of the world's oceans. In
spite of its small size, the Mediterranean is uniquely
important in the modern world, not only to the economic and
cultural life of its 18 coastal states,but as a contact zone
between Europe, Africa and Asia, and as a prime artery for
world trade via the Suez Canal. The states of the Mediterranean are characterised by immense ethnic, cultural and
religious diversity, and it is a region where historically,
violence has been endemic. On the face of it, the forces for
division are far stronger than those for harmony, and it is
therefore noteworthy that in the past decade there has been
considerable progress towards cooperation over certain environmental concerns.

This paper explores the scope for cooperation and the
potential for conflict in the Mediterranean in three important areas: a) fishing, b) seabed mining, and c) pollution control. Offshore oil and natural gas are considered
in a separate chapter, but it is worth noting that there are
differences between energy resources and the activities
being considered here because oil concession areas have to
be precisely located and the gain or loss of a few square
kilometres of seabed can be quantified in terms of potential
revenue. The activities considered in this paper are of
interest to states over a longer period of time, and the

41

dangers of conflict are less likely to be over boundary lines, than matters of principle and law.

Not all "Mediterranean" states have the same degree of interest in the Mediterranean Sea. There is a tendency to expect uniform commitment from all coastal states of the Mediterranean to matters concerned with the sea, but this is not realistic. Table 2.1 gives some crude indication of the importance of a Mediterranean coastline to 18 states, by calculating the number of square kilometres to every kilometre of coastline. A more interesting calculation might be the proportion of the population of each state resident within 50 km (or any other reasonable distance) of the sea. This would undoubtedly show results quite unlike those in Table 2.1, with Malta, Cyprus, Monaco and Libya high in the ranking and France, Spain, Turkey and Yugoslavia low. Table 2.2 is another useful measure, based on the hypothetical maritime boundaries shown in Figure 2.1, indicating approximate areas of seabed allocated to each coastal state. The actual delimitation, when it occurs, will doubtless differ from Figure 2.1 in a number of respects, but the ranking of states is probably about right. There are at least 31 potential international boundaries in the Mediterranean, only five of which have been agreed.

Discussion in this paper is confined to the coastal states, but it should be noted that non-coastal states have significant interests in the Mediterranean. It is a vitally important shipping route for Suez Canal traffic, and for the Black Sea states (including the Soviet navy). The United Kingdom has relic colonial interests in Gibraltar and Cyprus (where the Sovereign Bases could -perhaps - be considered eligible for a share of the seabed). Moreover, several landlocked states will "have the right to participate on an equitable basis in the exploitation of an appropriate part of the surplus of the living resources of the exclusive economic zones of adjoining states", (1) if the Treaty on the Law of the Sea is finally ratified. This could happen within four or five years. Although Article 69 of the treaty stipulates that "the relevant economic and geographical circumstances of all the states concerned" would be taken into account, and that developed landlocked states may only exercise their rights in the waters of adjoining developed states, there is some possibility of hard bargaining on this issue in the future.

Figure 2.1   Maritime boundary status (1982)

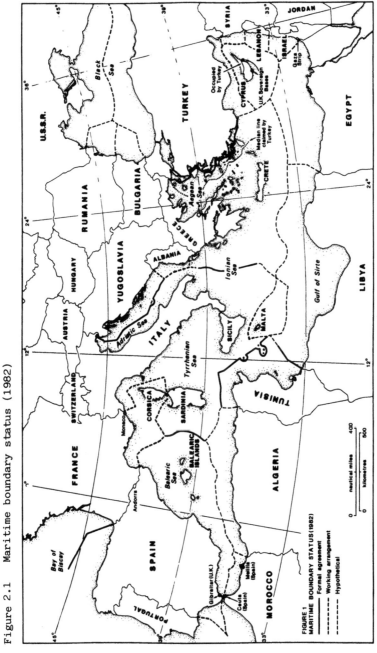

Table 2.1: The importance of Mediterranean coastlines to Mediterranean states

		Length of coastline (km)	Land area (km^2) per km of Mediterranean coast	Mediterranean coast as percentage of all coasts
1	Monaco	5	0.40	100
2	Malta	93	3.36	100
3	Cyprus	537	17.23	100
4	Greece	3,048	43.50	100
5	Lebanon	195	53.13	100
6	Italy*	4,539	66.36	100
7	Israel	222	93.33	96
8	Albania	284	101.22	100
9	Tunisia	1,028	159.73	100
10	Yugo-slavia	789	324.32	100
11	Turkey	1,804	424.96	51
12	Spain**	1,146	440.71	54
13	France***	910	606.23	36
14	Egypt	996	1,004.27	41
15	Libya	1,685	1,043.68	100
16	Morocco	352	1,162.50	21
17	Syria	152	1,226.84	100
18	Algeria	1,104	2,228.71	100

*   includes Sardinia and Sicily
**   includes Balearics
***   includes Corsica

Sources: various

Table 2.2: Allocations of Mediterranean seabed according to hypothetical boundaries in Figure 1

Country	Seabed area (km^2)
Italy	532,500
Greece	460,500
Libya	320,500
Spain	252,500
Egypt	176,500
Algeria	111,500
Tunisia	97,500
Cyprus	89,500
France and Monaco	88,570
Yugoslavia	70,250
Malta	61,190
Turkey	59,500
Morocco	21,750
Albania	21,000
Israel	20,500
Lebanon	15,750
Syria	10,250
TOTAL	2,409,760

Source: Measurements taken from Figure 1; the discrepancy between the area shown above and the true area of the Mediterranean (2,511,000 km) is due to the problem of measuring area precisely on Figure 1 which lacks the detail of large scale maps.

Table 2.3: Mediterranean fish catches by country (tons)

	1968	1978
Albania	4000	4000
Algeria	18200	34143
Bulgaria	1015	11
Cyprus	1354	1245
Egypt	13560	11636
France	39587	42400
Gaza Strip	3676	4700
Greece	55815	69753
Israel	3288	3550
Italy	296952	336699
Japan	–	86
Lebanon	2500	2400
Libya	5000	4803
Malta	1300	1044
Monaco	–	–
Morocco	10578	31410
Romania	1	–
Spain	83197	153876
Syria	800	1361
Tunisia	14537	35665
Turkey	33287	9290
USSR	100	–
Yugoslavia	30061	37464
TOTAL	618,809	786,535

Source: General Fisheries Council for the Mediterranean, Statistical Bulletin No. 3, Nominal Catches 1968–78, FAO, Rome, 1980, pp. 1–9.

FISHING

Importance of Mediterranean Fisheries

By world standards, Mediterranean fish stocks are
small largely because of the relatively unfavourable conditions for the development of fish food. The rate of phytoplankton production is a major factor in determining the abun-
dance of fish since plankton is the basis of the fish food
chain. Plankton production in the Mediterranean is inhibited
by scarcity of phosphates, nitrates and nitrites which are
essential nutrients. (2) There are few major rivers to bring
down nutrients and the building of the Aswan dam has greatly
reduced the input of nutrients from the Nile. Mediterranean
water is also relatively stable and little mixing occurs
between the surface and nutrients found at depth. Phytoplankton is most abundant in the north and west of the
Mediterranean and in the Adriatic Sea, and declines towards
the south and east. (3)

Environmental limitations mean that the Mediterranean
will never sustain a large-scale fishing industry. Fish
catches in the Mediterranean in 1978 totalled 786,535 tons,
or about one per cent of the world catch (Table 2.3). On the
other hand, Mediterranean fisheries are notable for the
great variety of species fished commercially (about 120 out
of a total of 500 species), and for the unusually high value
of the catch. The reasons for this are high demand
associated with culture and tradition, and reinforced
seasonally by the demands of the tourist trade. Fish prices
are probably still among the highest in the world, as was
the case in 1970 (Table 2.4). Locally, coastal fishing is
extremely important to the economy of coastal communities,
with small boats using traditional techniques being responsible for a high proportion of the catch. Many fishermen in
the Mediterranean are part-time. Scores of colourful fishing
boats add greatly to the scenery in the small ports and
harbours and perhaps reinforce the illusion of the importance of fishing to the national economies.

Mediterranean fish catches are already insufficient to
meet local demand. Between 60 and 70 per cent of fish
consumed in the region comes from other seas. France,
Morocco and Spain obtain a high proportion of their fish
from the Atlantic, while Turkey obtains a high proportion
from the Black Sea. In addition, Egypt, Greece, Israel and
Italy have ocean-going fleets while there are plans in

Table 2.4: Average unit value of fish landings 1970 (US$ per ton)

Mediterranean Sea	561
Indian Ocean	250
Pacific Ocean	94
Atlantic Ocean	126

Source: United Nations FAO, "Fish resources in the Mediterranean and Black Seas", Consultation on the Protection of Living Resources and Fisheries from Pollution in the Mediterranean, pp. 1-22, Rome, 1974.

several other countries, including Libya and Tunisia, to develop such fleets. By 1985 the demand for fish could reach anything from 4,350,000 to 6,100,000 tons, (4) but only a fraction of the extra demand is likely to come from the Mediterranean. The exact size of the unexploited surplus stocks in the Mediterranean is debatable. Certain species are probably overfished already. There appears to be some potential for increasing pelagic (surface swimming) catches in the eastern Mediterranean (5) and demersal (bottom dwelling) catches along the southern and eastern coasts of the eastern Mediterranean. There are unlikely to be surplus stocks in the heavily fished northern waters of the Mediterranean, where demersal species are already overexploited. (6) The best forecasts suggest that present catches could be doubled in volume, though the make-up of the catch will change. (7) This prospect is sufficiently attractive to make investment in the fishing industry worthwhile for several countries. At the same time rising demand will create higher prices and more powerful incentives to overfish. Competition between fishermen is bound to increase. A factor to be borne in mind is that some Mediterranean ocean-going vessels are going to be denied access to non-Mediterranean grounds with the likely introduction globally of exclusive economic zones to 200 nautical miles.

Some pressure could be taken off marine fisheries by the adoption of acquaculture at various inland locations. The potential for production of fish, shellfish, and molluscs by this method is high. (8) On the other hand pollution is already causing a decline in fish stocks in certain coastal locations and this could offset new developments in acquaculture if unchecked. Near the sewage outfalls of many large cities the presence of excess nutrients in the water has resulted in a phenomenon known as "phytoplankton bloom" in which unconsumed phtyoplankton decays and causes oxygen depletion in the water and the destruction of other forms of marine life. Elsewhere fish stocks have declined due to the pollution of coastal posidonia seagrass beds which are important for fish spawning and feeding. The effects of pollution on fish stocks at the regional level is not known, but it may not yet be very significant.

The marine fauna of the Mediterranean is closely related to the subtropical Atlantic, though certain Red Sea species have migrated to the eastern Mediterranean since the opening of the Suez Canal in 1869. Many Mediterranean fish are shallow water dwellers which tend to be caught close to the shore, including some of the demersal species. Pelagic fish congregate in shallow waters where plankton is abundant. Moreover, in summer they migrate inshore in response to coastal upwelling and their fishing tends to be seasonal. As a rule, therefore, much Mediterranean fishing is conducted within areas where the continental shelf is less than 200 metres deep. Thus disputed fishing grounds have tended to be in continental shelf areas not far from the coast. Table 2.5 shows some of the largest catches by species for the Black Sea and Mediterranean Sea in 1977.

Exclusive Fishing Claims and Disputes

As in other oceans of the world where fishing is important, Mediterranean states have made varying claims to exclusive fishing zones in recent years, though only four are beyond the limits of territorial waters (Table 2.6). International disputes arising directly as a result of national claims to exclusive fisheries have been rare in the past. A high proportion of fish catches are obtained within sight of the shore, or close to it, and few fishing vessels made their way to foreign grounds except by consent (the Greek and Italian sponge fishermen off Libya for example). In recent years however some international disputes have occurred between Mediterranean states, by far the most serious being that between Spain and Morocco over Atlantic fisheries. Faced with declining catches in 1973 Morocco declared an extension of exclusive fisheries from 12 to 70 nautical miles. Spain objected, and in spite of various efforts to reach agreement differences have not been

Non-energy Resources

Table 2.5: The most important species by volume in Mediterranean and Black Sea catches in 1977

Species or groups of species	Catches in tons
Pelagic Finfish	
Anchovy (Engraulis encrasicolus)	352,736
Sardine (Sardina pilchardus)	184,243
Horse mackerels (Trachurus spp)	56,937
Demersal Finfish	
Red mullets (Mullus spp)	21,100
Bogue (Boops boops)	27,301
Hake (Merluccius merluccius)	24,346
Picarels (Maena spp)	16,350
Grey mullets (Mugilidae spp)	14,237
Molluscs and crustaceans	
Mussels (Mytilus spp)	11,200
Cuttlefish (Sepia officinalis)	10,699
Shrimps, prawns	13,800
Common octopus (Octopus volgaris)	10,033

Source: United Nations FAO, Yearbook of Fishery Sta-tistics, Vol. 44, 1977, Rome 1978, pp.205-206.

resolved. The situation is complicated by the activities of Polisario guerrillas who arrested Spanish and Portuguese vessels in 1981, and the subsequent recognition of Polisario by Spain and Portugal. (9) Though the Spain-Morocco dispute concerns the Atlantic it could affect the Mediterranean particularly in view of Spain's retention of the enclaves of Ceuta and Melilla. In February 1979, for example, two Spanish fishing boats were arrested by the Moroccans of Melilla at a time when the Atlantic fishing dispute was erupting once again. (10)

The introduction of a 200-nautical-mile exclusive economic zone (EEZ),as proposed in the UN Treaty on the Law of the Sea, could create friction between Mediterranean states, though the implications of the EEZ have been foreseen for many years and it would be surprising if matters cannot be resolved peacefully. Only about 4 per cent of

Table 2.6: Mediterranean Sea: Territorial water claims and exclusive fishery claims (nautical miles)

State	Territorial water claim		Exclusive fishing claim	
	Distance	Year	Distance	Year
Albania	15	1976	Same as T.W.	
Algeria	12	1963	Same as T.W.	
Cyprus	12	1964	Same as T.W.	
Egypt	12	1958	Same as T.W.	
France	12	1971	Same as T.W.	
Greece	6	1936	Same as T.W.	
Israel	6	1956	Same as T.W.	
Italy	12	1974	Same as T.W.	
Lebanon	Undeclared	-	6	1921
Libya	12	1959	Same as T.W.	
Malta	12	1978	24	1978
Morocco	12	1973	200 (1)	1976
Monaco	12	?	Same as T.W.	
Spain	12	1977	Same as T.W.	
Syria	12	1964	Same as T.W.	
Tunisia	12	1973	Same as T.W.	
Turkey	6	1964	12	1964
Yugoslavia	12	1979	Same as T.W.	

(1) 6 miles in the Strait of Gibraltar

Source: The Statesman's Yearbook 1981-82 (J. Paxton, ed.), Macmillan 1981, XXV-XXVII.

Mediterranean fish catches come from the waters of other states. The introduction of the EEZ would deprive Spain, Greece and Italy of the right to fish in the waters off Algeria, Morocco, and Tunisia, though the loss would probably be "minor". (11) Japan would also lose access to certain tuna fisheries. Agreements could be reached between the parties for continued access to traditional fishing grounds, but it is a reasonable guess that the Maghreb countries will be in no hurry to negotiate such arrangements. A "stress zone" could therefore emerge in the western Mediterranean along the median line between Europe and Africa. Pending adjudication by the International Court of Justice on the disputed boundary between Malta and Libya this "stress zone" may be considered to extend further east.

Fishery disputes, rather like boundary disputes, are often symbolic of poor political relationships between states. This being so, the eastern Mediterranean might be expected to see many fishing quarrels in the future. Fishing is however relatively unimportant in the eastern Mediterranean and international conflict is unlikely over a matter which features so little in the national economies (Table 2.7). The catch for the Levant is about three per cent of the Mediterranean total, while that for the Aegean is only seven per cent. The Aegean could become the scene of a dispute if Greece chose to implement a 200-mile EEZ too hastily. A proportion of Turkey's modest Aegean catch undoubtedly comes from waters round the Greek Islands, and there are surplus stocks of pelagic fish still to be had. Matters could be complicated with the extension of European Community (EC) common fisheries policies to Greek waters. The exact shape of this policy is still unclear, but it could mean a 200-mile exclusive fishing limit as elsewhere, in which other Community members might seek to acquire the right to extract a quota of fish. In practice this is improbable because the Aegean is not a traditional ground for any other member state.

EC waters already extend over the coastal waters of France and Italy and will extend still further when Spain joins in 1984. The application of the 200-mile EC fisheries policy would place much of the western Mediterranean in Community hands. France, Italy and Spain are likely to retain exclusive control over six nautical miles, and partial control over the six- to twelve-mile zone, beyond which Community states would have free access. There is clearly considerable scope for argument and litigation between Community members in the western Mediterranean, but such disputes are likely to be confined to verbal exchanges. In a recent case, for example, the European Court of Justice ruled that a Spanish fisherman was within his rights to fish in France's six- to twelve-mile zone (off Bayonne in the Bay of Biscay) on the grounds that EC regulations cannot supercede an earlier (1967) Franco-Spanish agreement. (12)

Table 2.7: Fish catches in the eastern Mediterranean in 1978 (tons)

Levant		Aegean	
Cyprus	1,143	Turkey	5,555
Egypt	11,619	Greece	62,965
Gaza Strip	4,700	Total:	68,520
Israel	3,350		
Lebanon	2,200		
Syria	1,252		
Turkey	1,788		
Total:	26,052		

Source: United Nations FAO, General Fisheries Council for the Mediterranean, Nominal Catches 1968-78, Rome, 1980, pp. 7-8.

Conclusion - Fishing

There will undoubtedly be disagreements between Mediterranean states over fishing rights in the next few years. Declining stocks of certain species and rising prices will ensure a high level of government concern over fisheries. With the introduction of new maritime boundaries and the growth of complex EC regulations many of the disputes will simply be wrangles over quotas and legal matters. The delimitation of new boundaries could equally give rise to disputes which, given a poor political environment, could deteriorate rapidly. Such disputes are however unlikely to erupt into the kind of armed conflict which might destabilise the whole region. Apart from Spain and Morocco (in the Atlantic) only two fishing disputes in recent years appear to have involved shooting (Italy and Tunisia in 1975 and Albania and Yugoslavia in 1976). There were serious incidents involving Sicilian fishermen and the Tunisian authorities again in September 1982 when at least 16 Italian boats were arrested. (13) While fishing is of considerable importance to local communities it is not an activity which governments are likely to defend at all costs (Table 2.8).

Table 2.8: Fish and fish products by value in 1979 (thousands of US dollars)

	Value	Percent total exports
Egypt	750	0.04
France	26,103	0.26
Greece	13,445	0.34
Israel	30,805	0.67
Italy	12,125	0.16
Lebanon	440	0.05
Turkey	18,500	0.81
Tunisia	17,046	0.95
Spain	410,390	2.29
Yugoslavia	30,298	0.44
Others	Negligible	Negligible

Source: FAO, Trade Yearbook, Vol. 34, 1980.

The work of the General Fisheries Council for the Mediterranean (GFCM) is encouraging evidence of the desire for cooperation rather than conflict. The GFCM is an inter-governmental body set up within the framework of the FAO in 1952. In the past thirty years the work of the GFCM has greatly increased, and membership has grown to include all Mediterranean states (except Albania), plus Bulgaria and Romania. Reports of the GFCM sessions (held every two years) give an indication of the high level of cooperation between states, and a wide range of practical activity is engaged in by the GFCM. There is now widespread recognition that the problem of Mediterranean fisheries must be tackled regionally, and at governmental level. Examples of the acti-vities of the GFCM are: a) collection and analysis of data on the present condition of stocks; b) recommending and implementing measures for the conservation and rational management of resources (e.g. by regulating methods, periods and areas where fishing is permitted, regulating catch sizes, fishing efforts, etc.); c) research into the effects of pollution on marine life; d) encouragement of alter-native forms of fish production (e.g. through acquaculture and use of brackish water); e) training of fisheries personnel; f) promotion of advances in fisheries technology in a variety of fields. (14)

While it would be easy to find many defects in the work of the GFCM, particularly with respect to enforcement, the significant point is that it has survived and grown, and most Mediterranean states have demonstrated their desire for practical cooperation through the GFCM. International cooperation of this kind is no insurance against disagreements, but as such work progresses the less likely conflict becomes. For example, the GFCM has sought to limit the fishing effort of member states to avoid overfishing. France and Spain have introduced a licensing system limiting the number and type of new trawlers while Greece, Italy, Tunisia, and Yugoslavia have curtailed credit for building new trawlers. (15)

SEABED MINING

Introduction

A great deal of publicity has been given to the prospects for the recovery of minerals from the seabed. There is a growing volume of scientific literature on the subject, and the necessary technology has already been developed. It is an assumption in the United Nations draft Treaty on the Law of the Sea, in which 50 articles are devoted to seabed mining, that it is desirable and inevitable. During the latter sessions of the Third UN Conference on the Law of the Sea (UNCLOS) which evolved the proposed treaty, arrangements for mining the seabed beyond the 200-mile limit were hotly debated. In April 1982 the United States finally voted against the treaty on the grounds that it would restrict freedom of access to seabed minerals by United States companies. Opinions differ concerning the economics of mining the seabed, but the probability is that land-based sources will be cheaper at least until they near exhaustion. In a world where economics had the final say, deep seabed minerals would probably be left alone for at least another twenty years, but there is a desire to extract urgently for strategic reasons. The United States and NATO depend very heavily on imported supplies of certain key minerals which could be obtained from offshore sources.

The argument in the Third UNCLOS has been over control

of seabed minerals beyond the 200-mile limits. Broadly speaking, the Third World states want maximum international control so that this new source of wealth will benefit the poor countries. They also fear the effect on the price of land-based minerals if the advanced industrial countries were able to supply themselves from the oceans. The draft treaty proposes an International Seabed Authority to administer seabed minerals. The headquarters of the ISBA are to be in Jamaica, though Malta was very nearly chosen. In some ways Malta would have been an inappropriate location since the Mediterranean Sea is unlikely to have any seabed left as "the common heritage of mankind" once the 200-mile limits have been drawn up (Figure 2.1). The management and exploitation of Mediterranean seabed minerals will thus be the exclusive right of the coastal states, if exploitation proves desirable. The Mediterranean can be regarded as a microcosm, however, of certain aspects of the global debate about minerals, notably the strategic argument for exploitation and the conflict of interest between Third World mineral producers and the industrial world consumers. Although it sounds an attractive proposition, the possibility of declaring certain areas of the Mediterranean seabed (below 2,000 metres for example) as common heritage does not seem realistic in view of the extent of existing national claims and expectations.

Mediterranean Seabed Resources

Offshore mineral resources in the Mediterranean fall into four categories: evaporites, placers, aggregates and metalliferous muds.

Evaporites. Deep drilling in the Mediterranean seabed in 1970 revealed an extensive layer of evaporites reaching up to 3,000 feet thick in places, probably some five million years old. They were formed by the drying out and flooding of a former Mediterranean Sea many times over long periods. These evaporites consist of rock salt, sulfur and potash salts in huge quantities: in Sicily and certain other Mediterranean islands they reach the surface and are mined as the basic raw materials of the chemical industry. The commercial exploitation of seabed evaporites however is not regarded as currently feasible. Apart from the problems of working in an underwater environment the deposits are overlain by thick marine sediments. At the same time reserves of most of the constituents of the evaporites are

56

in plentiful supply on land and readily accessible. Their relevance to the discussion of offshore resources lies largely in their very long-term potential. When maritime boundaries which could last for many decades are being drawn up, statesmen and lawyers have to bear in mind unforeseen advances in technology which could turn a reserve into a resource. Evaporites will not cause conflict between states, even though they tend to be located in central areas of the seabed (16) where international boundaries will run, but they could be one reason among many why coastal states may seek to negotiate maximum allocations of the seabed. Evaporites in the shallower coastal waters could conceivably be of some economic value in the distant future.

Placer deposits. Placer deposits are found in certain coastal locations where metal-bearing rocks on land have been weathered and the debris washed out to sea by rivers. The heavy metal particles accumulate as the action of waves, currents and tides removes lighter materials. Placers may be found either on beaches above water level, on submerged ancient beaches, or in buried river valleys and depressions on the sea floor. In all cases they tend to be close to the coast, on the continental shelf. The only possible disputes might be on the boundary between adjacent states, but these are considered highly improbable. Placer deposits are neither sufficientlyextensive nor of sufficient commercial value to be of political importance. If exploited, the returns are likely to be rather small. (17) The only known mining of placers in the Mediterranean occurs along the Egyptian Nile Delta coast where iron, tin, zirconium, titanium and monazite are obtained from extensive mineral sands washed down by the Nile. The dredging of these minerals is clearly not a source of international conflict. Placers have been reported quite widely elsewhere in the Mediterranean, though their economic potential is not clear. The most important are probably iron (Italy, west coast; Greece, north-east coast; Tunisia, north coast; Algeria); titanium (Corsica, east coast; Greece, north-east coast); and chromite (Yugoslavia; Greece; Turkey, west coast; Cyprus, north-east coast). Italy's Sardinian shelf may be particularly promising for rutile, titano-magnetites, ilmenite, zircon, cassiterite, monazite and cobalt. (18) There are also deposits of titanium and valuable phosphates on Morocco's Atlantic coast. Phosphates are also found on

the continental shelves of Malta, Spain and Italy. At some
future date it could prove attractive to exploit some of the
richer coastal deposits of chromite and titanium, but it is
difficult to foresee circumstances in which their exploitation would lead to serious conflict between states. Greece
or Cyprus might conceivably object to any possible Turkish
move to dredge chromite from the north-east coast of
Turkish-occupied Cyprus, but there is no evidence that such
exploitation has ever been contemplated. According to one
authority, titanium is likely to be one of the earliest
placer deposits worth exploiting on the continental shelf.
(19)

   Metal-bearing aggregates and crusts. The Mediterranean sea has no major seabed deposits, largely because
rates of sedimentation are too great to permit the formation
of manganese nodules, as found in all the other oceans of
the world, the most notable concentrations being in the
Pacific and Atlantic Oceans. There are however more than 40
volcanoes in the Mediterranean, and in recent years geo-
logists have become interested in the metal-bearing crusts
found associated with them. Since they lie in relatively
shallow water (500-1500 metres), many of them close to the
shore, they have been regarded as a potential source of
"industrially interesting" raw materials. (20) Mineralisa-
tion which forms veins and pockets within a silty-clayey
material consists chiefly of manganese and iron. In the
Tyrrhenian Sea the manganese content is around 48 per cent,
but in the Aegean more iron (20-40 per cent) is present.
(21) Much thought has been given to the origins of these
mineral deposits, particularly by Italian scientists, the
suggestion being that they were derived as hydrothermal
solutions emanating from current-swept volcanic crests. More
work will be necessary before all the Mediterranean
volcanoes have been investigated, but under present world
economic conditions the relatively small amounts of manganese may not be commercially worth recovering. On the
other hand, the shallowness of the water and the prospects
of recovering nickel, cobalt and copper have raised hopes of
exploitation in recent months. (22) In time of war, or in
the distant future when manganese nodules have become unob-
tainable (for example, because of quarrels within the ISBA),
Mediterranean countries might turn to their own more costly
local resources. It is not known which are the most promising sites, but a preliminary attempt to plot volcanoes and

Figure 2.2   Submarine volcanoes

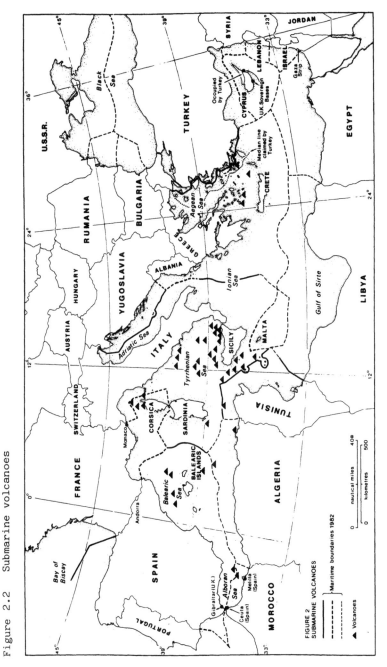

maritime boundaries (actual and possible) reveals surprisingly few potential disputes (Figure 2.2). The most
obvious difficulty could be in the Alboran Sea between
Morocco and Spain, where disputes over fishing, the status
of Ceuta and Melilla, and the effect of Alboran on maritime
boundaries already create a zone of potential conflict. On
the other hand, the volcanic aggregates are not regarded as
a probable source of conflict, except in conjunction with
other disputes. It will be noted in Figure 2.2 that almost
all the volcanoes fall within the jurisdiction of the industrial countries of the northern Mediterranan.

Metalliferous muds. Italian and French research teams
are investigating the potential of metal-bearing muds analogous to the well-known deposits of the Red Sea. Zinc,
silver, lead, copper, cadmium, gold and cobalt could be
present. The most promising areas are to the south of Rhodes
and Crete, in the basins of Cyprus and in the system of arcs
in the Aegean. (23) Commercial exploitation of these muds is
a remote possibility, and political problems associated with
their exploitation should not occur, with the possible exception of the Cyprus basins.

Conclusion - Seabed Mining

If Mediterranean seabed minerals are exploited in the
future it is likely to be for strategic reasons, with the
industrial north taking the lead. Two minerals which might
be sought early are manganese from volcanic sites and
chromite from placer deposits. Manganese is an essential
ingredient in alloy steels and is used in producing dry cell
batteries. Present world production and reserves are dominated by South Africa and the Communist bloc. Since manganese is regarded as a strategic mineral, United States and
NATO dependence on imports is regarded as undesirable by
defence experts. Chrome is even more important
strategically, for use in the production of high technology
alloys, armour plate and other defence needs. It is resis-
tant to corrosion and oxidation. NATO is heavily dependent
upon imports, with South Africa, Turkey and Zimbabwe as the
chief sources. South Africa and Zimbabwe together hold 95
per cent of reserves. In these circumstances local sources
of manganese and chrome will be seriously investigated. As
physical depletion of land-based supplies draws near in the
next century there could be an economic case for exploitation.

Any moves to exploit Mediterranean minerals could meet with resistance from the Mediterranean countries which are currently exporters of the same minerals. This is more likely to be a source of regional friction than a question of boundaries and ownership. In particular, Turkey has an interest in the world chrome market (293,000 tons in 1978) and Morocco in manganese (135,700 tons in 1979) and iron ore (61,700 tons in 1979). Tunisia is also an exporter of iron ore (1977 production: 343,000 tons). (24) As a footnote, it may be that offshore mineral exploitation will create unacceptable environmental problems through turbulence and the deposition of tailings, etc. This could be a further source of friction between states.

ENVIRONMENTAL PROTECTION

Introduction
The physical characteristics of the Mediterranean Sea which render it particularly vulnerable to environmental damage through pollution can be summarised as follows: a) The surface area and volume of water are not great in relation to the coastal population densities; b) the seawater is relatively stable. The only major exchange of water occurs via the narrow strait of Gibraltar. It takes some 80 years to change the water completely; c) evaporation rates are high and exceed the input of freshwater by rainfall and river flow about three times; d) there are numerous islands and long coastlines (about 20,000 km.); e) tidal regimes and coastal currents are often weak and their 'scouring' effect is limited. Tides are generally from two to four metres.

Human activity during the past two or three decades has brought the Mediterranean marine environment to a point of crisis. The fundamental causes of this growth of pollution may be summarised briefly as follows: a) rapid growth of coastal populations by migration and natural increase, with a high proportion concentrated in a few large cities. Coastal populations number over 60 million; b) massive seasonal tourism generated largely in Europe which doubles the resident coastal population; c) high rates of industrial growth, notably in the countries of the north with emphasis on coastal locations for such polluting industries as oil refining, iron and steel, petrochemicals and cement; d) .

the use of the Mediterranean for the transportation of petroleum and petroleum products and other seaborne cargoes.

In the following paragraphs, it is assumed that all these activities will grow in volume over the next few years, some of them very substantially. The need to exercise increasing controls to protect the environment from further damage is well recognised, but the implementation of these controls offers considerable potential for disagreement as well as cooperation between states.

Types of Pollution

The nature of Mediterranean pollution has been admirably diagnosed elsewhere, and is discussed in this paper only to identify likely causes of conflict.

Industrial and chemical waste. Industrial and chemical pollution reaches Mediterranean waters through river flow, in the air, and directly from coastal industrial activities. It was only ten years ago that the first authoritative investigation of land-based pollution in the Mediterranean was completed. (25) Much research has been conducted since, and the problem clearly remains serious. This category of pollution is largely the product of coastal iron and steel making, oil refining, petrochemicals and other chemical industries. It also includes certain pollutants such as pesticides derived from agricultural activity. The distribution of industrial pollution is uneven, with France, Italy and Spain being responsible for the largest quantities, the Po and the Rhone being the most offending rivers. At one time the Rhone was alleged to release each year "30,000 tons of petroleum hydrocarbons, 500 tons of pesticides, 700 tons of phenols, 1,250 tons of detergents and 24,000 tons of organic pollution". (26) Parts of the Mediterranean far from the shore are still relatively free from this kind of pollution. The worst concentrations are found locally, usually near the outfalls of major industrial plants. Industrial pollution, however, must be the concern of all the coastal states of the region since its major effect is upon marine life which recognises no international boundaries. Marine organisms can concentrate metals (e.g. mercury, cadmium, copper, zinc, lead) from the water and these endanger human health when contaminated sea food is eaten. Many cases have been reported in the Mediterranan.

Domestic sewage. The quantity of sewage reaching the Mediterranean is colossal on account of the dense coastal populations of residents and visitors. According to one report at least 120 coastal towns and cities still pump 90 per cent of their sewage into the sea untreated or poorly treated, (27) which seems little better than ten years ago. Treatment plants are being constructed urgently, but in many cases the need to cope with seasonal influxes of tourists has led to unexpectedly high costs. Sewage in coastal areas is aesthetically unpleasant and a threat to the tourist trade, but it is also a serious hazard to health. There are numerous cases all round the Mediterranean of beaches having to be closed for a time because of sewage pollution. In Spain, for example, 37 per cent of all beaches were found to fall below the minimum standards set by the World Health Organization (WHO) in 1981. (28) Illnesses associated with infection resulting from bathing in contaminated water or eating contaminated seafood (e.g. oysters, mussels) are common in the Mediterranean. Cholera, dysentry, typhoid and hepatitis have been associated with marine pollution resulting from sewage carrying pathogenic bacteria and viruses. The claim that someone swimming in the Mediterranean has a "one in seven chance" of catching a disease dramatised the dangers a few years ago, but the risks are still much the same. (29)

Sewage pollution tends to be confined to coastal locations and to the vicinity of major towns. It does not have great political significance, though it is almost certain that somewhere in the Mediterranean there are beaches allegedly polluted by the sewage from a neighbouring town on the other side of an international boundary.

Oil pollution. The Mediterranean is more seriously polluted by oil than any other ocean; about half the floating oil in the world is in the Mediterranean Sea, and in places 500 litres per sq.km. have been found. (30) The total quantity entering the sea in a year has been calculated from 0.5 to one million tons (31) compared with six million tons for the world. This astonishing quantity is likely to increase very substantially in the future unless conservation measures implemented internationally are highly successful. The achievement of effective oil pollution control could prove to be a serious source of international controversy. The danger of increased oil pollution is likely for several reasons:

1) Increased volumes of oil moving across the Mediterranean to Western markets as a result of increased pipeline

capacity, and a possible further widening of the Suez Canal.
Table 1.6 indicates the sources of this oil, even without a
widening of the Suez Canal.

2) More spills resulting from increasing oilrefining
activities in coastal locations, above all in the north.
The total quantity could rise at worst from 10,000 tons in
1978 to 13,200 tons per annum in 1985. (32)

3) The likelihood of a larger number of accidents
which have been remarkably few in the Mediterranean hitherto, as a direct result of heavier tanker traffic. (33)

Detailed statistics are unavailable as to the sources
of oil pollution in the Mediterranean, but estimates for the
world as a whole give some useful indicators (Table 2.9).
More than 40 per cent is associated with the transportation
of oil. This is highly relevant to the Mediterranean where
much oil is in transit from and to countries outside the
region. Apart from the obvious threat to tourism posed by
oil pollution, oil also constitutes a serious threat to
marine life. It is clear that certain parts of the Mediterranean are more at risk than others, with coasts in the
vicinity of the main tanker routes being particularly vulnerable.

Table 2.9: Sources of oil pollution in the world's oceans
(per cent)

Operational sources (cleaning, ballasting, etc.)	32.5
Accidents to tankers and pipelines	9.0
Offshore oilfield spillage	4.0
Land-based discharges	32.5
Natural seepage	15.0
Atmospheric rain-out	7.0

Source: P. Le Lourd, "Oil pollution in the Mediterranean",
Ambio Vol. VI, No. 6, 1977, p. 319.

Radioactive waste. Radioactive discharge into the Mediterranean does not appear to be important. France, Italy and
Spain have nuclear power plants which can be a source of
tritium and other radio nuclides, but so far there is no
evidence of major political controversy. Several other Medi-

terranean states are contemplating nuclear power, and regional monitoring and controls are on the agenda.

International Cooperation to Combat Mediterranean Pollution

The appalling implications of the growing pollution in the Mediterranean became clear in the early 1970s. There had previously been a number of measures taken by individual countries affecting their own waters and various international agreements which involved the Mediterranean to some degree, but concerted international action by Mediterranean states only began with the signing of the Barcelona Convention in 1976. This historic Convention arose out of the 1972 United Nations Conference on the Human Environment at Stockholm, which gave birth to the UN Environment Program (UNEP). In 1974 UNEP selected the Mediterranean for special action to protect the marine environment, no doubt partly as a result of an FAO consultation early in the year. (34) An Intergovernmental Meeting on the Protection of the Mediterranean was convened by UNEP in Barcelona in 1975. Cyprus and Albania were the only absentees. A Mediterranean Action Plan (MAP) was approved at Barcelona (35) and its implementation was formally agreed in the Barcelona Convention of 1976 which only Albania did not sign.

The Barcelona Convention. This timely agreement was an encouraging start to regional cooperation to combat pollution in the Mediterranean. It was basically an agreement to encourage national legislation, initiate regional planning and research, and make certain practical institutional and financial arrangements. It included a proposal to set up a Regional Oil Pollution Combatting Centre in Malta which is now operational. The Convention contains no sanctions and has no supranational authority. It has been accompanied by four important Protocols (36): a) Protocol for the Prevention of Pollution of the Mediterranean Sea by Dumping from Ships and Aircraft (Barcelona, February 1976); b) Protocol Concerning Cooperation in Combating Pollution of the Mediterranean Sea by Oil and other Harmful Substances in Cases of Emergency (Barcelona, February 1976); c) Protocol for the Protection of the Mediterranean Sea against Pollution from Land-based Sources (Athens, May 1980); (37) d) Protocol on Specially Protected Areas in the Mediterranean Sea (Geneva, March 1982).

Protocols (1) and (2) have been ratified by all Mediterranean states except Albania. Protocols (3) and (4) are still in the process of ratification, following initial approval by most Mediterranean states.

The Athens Protocol on land-based pollutants took over two years to negotiate partly because the European Community (EC) sought to ensure that it did not conflict with other Community legislation and agreements. The Protocol was approved by the EC and by all Mediterranean states except Albania, Syria and Egypt. Since about four-fifths of Mediterranean pollution is land-based, such an agreement was an extremely important step forward. The parties agreed to eliminate or strictly limit pollution reaching the sea directly, or through rivers, canals, or other watercourses, or in the atmosphere. A list of substances to be controlled is given in the form of a "Black List" (e.g. mercury, cadmium, persistent synthetic materials, radioactive substances) and a "Grey List" (e.g. zinc, copper, chrome, lead, pesticides, crude oils). Black list substances are to be eliminated and Grey list substances strictly controlled.

The Athens Protocol highlights some of the difficulties of pollution control. First, the riparian states of certain international rivers may have no direct interest in preventing pollution in the Mediterranean: for example, the Evros, Netos, and Strimon rivers which rise in Bulgaria and reach the Aegean through Greece. Similarly, Uganda, the Central African Republic, and Ethiopia are unlikely to exercise stringent controls on the upper stretches of the Nile. (38) Secondly, enforcement is left to the states themselves. Thirdly, there is the familiar conflict in the Mediterranean between environmental control, which is costly, and economic development. The major polluters are the developed countries of the north, while the developing states of North Africa create relatively little pollution. While this is partially recognised by the fact that France, Spain and Italy are likely to supply 85 per cent of the necessary funds ($15 billion over the next 15 years) (39), there are inevitably going to be some difficulties over deciding upon standards and guidelines agreeable to both north and south.

It is far too early to assess the effectiveness of the Athens Protocol. For some countries like France and Spain it will simply reinforce legislation already in existence. In other countries the effects can be expected to be rather slow. As with so much anti-pollution activity since the Barcelona Conference of 1975, one positive result will be the exchange of information and the production of large quantities of research data. The implementation of anti-pollution agreements will no doubt cause strains and stresses between states, but they at least have a common goal in sight from which everybody will benefit - a cleaner and safer Mediterranean.

Italy has also been instrumental in pioneering bilateral agreements with neighbouring states to combat pollution - with Yugoslavia in 1974 concerning the Adriatic, and with Greece in 1979 concerning the Ionian Sea.

The Blue Plan. The Blue Plan was proposed at a UNEP meeting in Nairobi in 1975, and after a series of consultations it was finally launched in September 1980. It is an attempt to balance the emphasis on pollution control with economic development. Its aim is to reconcile future development with environmental protection. There is no plan as such, but an emphasis on encouraging cooperation between states over common problems in general, not just those associated with the sea. Phase 1 involves the preparation of a series of expert studies on aspects of the Mediterranean region (e.g. industry, agriculture, migrations, energy, resources), each study being the responsibility of a joint north/south team. Phase 2 involves analysis of the Mediterranean as an integrated region, identifying common problems over the next three or four decades. Phase 3 will be the communication of recommendations to assist governments to achieve rational development plans from an environmental viewpoint. It would be a mistake to expect too much of the Blue Plan. Much will depend on the quality of the ideas it produces, which have yet to be seen. So far, there has been encouraging commitment to the concept, and only Cyprus, Albania and Libya have not participated in any meetings. It is a cautious but significant move towards practical cooperation, emphasising the one common concern which cuts across political differences - environmental protection.

Conclusion - Environmental Protection

Environmental protection has done much for inter-state cooperation in the Mediterranean since the early 1970s. In the next few years a reasonable degree of cooperation can be expected as more rigorous controls are exercised. The most likely cause of serious disputes between states will be oil tanker traffic. As the volume of oil transported increases, and awareness of the associated risks grows (possibly as a result of a major tanker disaster), states will be in-

67

creasingly keen to protect their waters from pollution. The proposed EEZ bestows on the coastal state the right to exercise jurisdication for "the preservation of the marine environment" (40), although other states are to enjoy the right of freedom of navigation. Thus the re-routing of tankers, or some control on their movement, could be in the interests of the coastal state. There could also be some resentment over the growing number of tankers in transit from the Suez Canal to non-Mediterranean locations. Egypt is the only Mediterranean state to benefit from this traffic in the form of canal dues. Malta, Tunisia, Algeria, Morocco and Spain on the other hand, all with coastal tourist resorts, could experience serious oil pollution in the event of a major accident. The Suez Canal was widened considerably between 1975 and 1980, enabling loaded tankers of up to 150,000 tons to transit, compared with the previous maximum of 60,000. The average size of tankers has thus already increased. But the Suez Canal could be widened still further to accommodate loaded supertankers of up to 250,000 tons. The project is still under discussion and a decision is expected shortly, but the implications for the Mediterranean are obvious if the scheme goes ahead. Table 2.10 indicates that a number of non-Mediterranean states have an interest in Suez Canal tanker traffic, and would no doubt welcome its widening still further.

CONCLUSION

This chapter has argued that there are no objective reasons why Mediterranean states should become involved in serious conflict in the areas of fishing, seabed mining, and pollution control. On the contrary, there are many oppor- tunities for international cooperation which would be in the interests of all the states concerned. This is not to say, however, that there will be no conflict in the future over these issues; the history of the Mediterranean shows how difficulties can easily arise between states over apparently minor matters, simply because they occur against a back- ground of generations of rivalry and suspicion.

Non-energy Resources

Table 2.10: Suez Canal: Northbound petroleum and products by unloading countries in 1981

	('000 tons)
Italy	12,939
Romania	5,154
Holland	2,117
Spain	1,845
Turkey	1,748
Morocco	1,587
USA	1,448
France	1,433
Greece	1,110
Egypt	1,065
Syria	884
United Kingdom	861
Portugal	743
Yugoslavia	497
Belgium	403
Denmark	247
Germany (Fed.)	239
Lebanon	194
Sweden	93
USSR	87
Canada	59
Gibraltar	50
Finland	47
Malta	30
Poland	30
Cyprus	28
Algeria	18
Others	1,610
TOTAL	36,566

Source: Suez Canal Authority, Monthly Report, Ismailia, December 1981, p. 66.

Non-energy Resources

Fishing

Increasing demands for Mediterranean fish are likely to
create dangers of overexploitation, and lead to more
national claims to exclusive fishing beyond present limits.
Local disputes over fishing rights and over levels of
catches of threatened species can be expected, but are
unlikely to be very serious at the inter-state level. Meanwhile, practical steps to manage Mediterranean stocks will
be extended through the GFCM, and clearly the scope for
cooperation over fisheries management is almost unlimited.

Potential trouble spots. Alboran Sea (Spain/Morocco/Gib-
raltar (UK); North African waters (Italy/Algeria/Tunisia);
Malta Channel (Italy/Malta); Libya/Egypt boundary zone;
Aegean Sea (Greece/Turkey).

Timescale. The 1980s could be difficult with the
introduction of EEZs, the accession of Spain and Greece to
the EC, declining stocks of certain species, and attempts by
the GFCM to exercise more control. Alternatives (e.g. aqua-
culture) will not be effective for some years.

Strategic implications. There is a north/south element
in potential disputes but these disputes are likely to be
short-lived or seasonal. Superpower intervention is out of
the question. Fishing disputes can, however, be used to
exacerbate deteriorating inter-state relations, and must be
handled with care.

Seabed Mining

Apart from submarine volcanoes, metalliferous muds, and
perhaps some placers, seabed mining sites in the Mediterranean are unimportant. There are some prospects for mining
manganese and iron from volcanoes and chrome from placers
but the scale is uncertain. They could possibly be used in
an emergency: there are enough sites away from maritime
boundaries to avoid conflict.

Potential trouble spots. In the unlikely event of
widespread exploitation, sites near boundaries (Figure 2.2)
might be disputed, and there could be arguments over
possible environmental effects on nearby coasts (e.g.
Spain/Morocco, Italy/Tunisia, Greece/Turkey). There could
also be resistance to offshore production from mineral
producers in the region.

Timescale. The most favourable seabed mining locations
(e.g. the Red Sea) are unlikely to yield minerals for at
least five years. The deep Pacific may not be exploited for

20 years or more. Similarly, Mediterranean sites are unlikely to be used for many years, with the possible
exception of volcanic sites in the Tyrrhenian Sea.

Strategic implications. Manganese and chrome are
strategic minerals which France and Italy might find
valuable if other sources were denied. The more remote
possibility of recovering cobalt should not be overlooked.

Environmental Protection

The willingness of 17 out of 18 coastal states to
cooperate over environmental protection from 1975 is encouraging. The first decade will be relatively easy (fact-
finding, preliminary planning, combating oil pollution,
etc.) but the next decade could reveal strains and stresses
between the industrial north and developing south. There may
also be differences over attitudes to oil tanker transit
traffic between those who benefit (major exporters and
importers) and those whose coasts are at risk. Danger areas
are along the main Suez Canal to Strait of Gibraltar route.
A major disaster (e.g. in the Strait of Gibraltar, off
Malta, or in the approaches to Suez) could change attitudes
radically).

Timescale. The next five years could reveal obstacles
to real progress in checking land-based pollution, followed
by some disillusionment. Tanker traffic will steadily increase in the same period. If the Suez Canal is widened
further it could be complete by the early 1990s.

Strategic implications. Several non-Mediterranean countries, including the USA and the USSR, receive oil imports
via the Mediterranean and would resist attempts to reduce
quantities or re-route tankers for environmental reasons.
Most importers would welcome a wider Suez Canal. Unlike
fisheries or pollution, states perceive energy supplies as a
potentially legitimate area for the use of force to protect
national interests.

NOTES

1. Third United Nations Conference on the Law of the Sea, Draft Convention on the Law of the Sea, August 1981, Article 69(1), p. 27.

2. See: D.A. McGill, Woods Hole Oceanographic Institution, Mediterranean Sea Atlas, Massachusetts, 1970, pp. 100-110.

3. See United Nations Food and Agriculture Organization (FAO), Atlas of the Living Resources of the Seas, Rome, 1972, Sheets 1.1 and C.1.

4. General Fisheries Council for the Mediterranean (GFCM), Perspectives on Fisheries Development to 1985, FAO, Rome, 1974, p. 74.

5. GFCM, Report of the Twelfth Session, FAO, Rome, 1974, pp. 7-9.

6. GFCM, Report of the Eleventh Session, FAO, Rome, 1972, p. 11. See also J.A. Gulland, The Fish Resources of the Ocean, FAO, Rome, 1971, pp. 39-41.

7. United Nations FAO, "Fish Resources in the Mediterranean and Black Sea", Consultation on the Protection of Living Resources and Fisheries from Pollution in the Mediterranean, Rome, 1974, p. 1.

8. D. Charbonnier, "Prospects for fisheries in the Mediterranean", Ambio, Vol. VI, No. 6, 1977, p. 376.

9. Keesing, Contemporary Archives, 27 April 1979, p. 29567; 13 February 1981, p. 30176.

10. Keesing, Contemporary Archives, 27 April 1979, p. 29567.

11. J. Gulland, "The new ocean regime: winners and losers", Ceres, Vol. 12, No. 4, 1979, pp. 19-23.

12. Telex Mediterranean, Brussels, No. 132, 9 February 1982, p. 16.

13. Financial Times, 8 September 1982, p. 6.

14. D. Charbonnier, op. cit.

15. GFCM, Report of the Fourteenth Session, FAO, Rome, 1978, p. 14.

16. The Atlas of the Oceans, Mitchell Beatley, London, 1977, pp. 140-141.

17. D.S. Cronan, Underwater Minerals, Academic Press, 1980, p. 295.

18. Antonio Brambati, "Some aspects of the mineral resources of the Mediterranean Sea", Lo Spettatore Internazionale, Vol. XVII, No. 4, 1982.

19. R.H. Charlier, "Other ocean resources", in E.M. Borgese and N. Ginsburg (eds.), Ocean Yearbook 1, University of Chicago, 1978, p. 183.

20. L. Morten et al., "Fe-Mn crusts from the southern Tyrrhenian Sea", Chemical Geology, Vol. 28, 1980, pp. 261-278; P.L. Rossi et al., "A manganese deposit from the South Tyrrhenian region", Oceanologica Acta, Vol. 3, No. 1, 1980, pp. 107-112.

21. E. Bonatti et al, "Submarine iron deposits from the Mediterranean Sea", in D. J. Stanley (ed), The Mediterranean Sea: a Natural Sedimentation Laboratory, Dowden, Hutchinson & Ross, 1972, pp. 701-710.

22. Antonio Brambati, op. cit. p. 2.

23. Ibid.

24. J. Paxton (ed), The Stateman's Yearbook 1981-82, Macmillan, 1981, pp. 1201, 875, 1193.

25. GFCM, The state of marine pollution in the Mediterranean and legislative controls, FAO, Rome, 1972.

26. Financial Times, 14 February 1978, p. 28.

27. The Guardian, 5 April 1982, p. 19.

28. The Times, 1 June 1982, p. 4.

29. Financial Times, 14 February 1978, p. 28.

30. P. Le Lourd, "Oil pollution in the Mediterranean", Ambio, Vol. VI, No. 6, 1977, p. 317.

31. Ibid, p. 319.

32. G. Luciani "Perspectives on industrialisation in the Mediterranean 1980-2000", Part II of Industrial growth and industrialisation strategies in the Mediterranean, IAI, Rome, 1981, Table 32.

33. S.C. Truver, The Strait of Gibraltar and the Mediterranean, Sijthoff and Noordhoff (Netherlands), 1980, p. 110.

34. United Nations FAO, Consultation on the Protection of Living Resources and Fisheries from Pollution in the Mediterranean, Rome, February 1974.

35. See E. M. Borgese and N. Ginsburg (eds), Ocean Yearbook 2, University of Chicago, 1980, pp. 153-182 and 547-554.

36. A.C. Lagrange "The Barcelona Convention and its Protocols", Ambio, Vol. VI, No. 6, 1977, pp. 328-332; see also: UNEP, Regional Seas Reports and Studies No. 18, Regional Seas Programme: Workplan, Geneva 1982.

37. See G. J. Timagenis, "Protocol for the protection of the Mediterranean sea against pollution from land-based sources", Hellenic Review of International Relations, Vol.

1, No. 1, 1980, pp. 123–136.

38. C. Odidi Okidi, <u>Regional control of Ocean Pollution: Legal and Institutional Problems and Perspectives</u>, Netherlands, 1978, p. 241.

39. <u>The Guardian</u>, 15 April 1982, p. 19.

40. Third United Nations Conference on the Law of the Sea, <u>Draft Convention on the Law of the Sea</u>, August 1981, Article 56(1), p. 21.

Part Two

LEGAL CONSIDERATIONS: Territorial Limits, Continental   shelf
and the Law of the Sea

Chapter Three

EXTENSION AND DELIMITATION OF NATIONAL SEA BOUNDARIES IN THE
MEDITERRANEAN
Geoffrey Marston

THE PHYSICAL BACKGROUND

Eighteen sovereign entities have coasts on the Mediterranean. They are Morocco, Algeria, Tunisia, Libya, Egypt,
Israel, Syria, Cyprus, Turkey, Greece, Albania, Yugoslavia,
Italy, Malta, France, Monaco and Spain. In addition, a nineteenth sovereign entity, the United Kingdom, has two dependent territories in the Mediterranean, namely Gibraltar and
the Sovereign Base Areas on the island of Cyprus.

The Mediterranean Sea extends for about 3,700 km. east-
ward from the Strait of Gibraltar to the coast of Syria and
has an area of 2,500,000 sq.km. or more than eight times the
size of Italy. It consists of two distinct parts separated
by a submarine platform running between Sicily and North
Africa.

The western part, the smaller of the two, can be
divided into several distinct areas: a) the Balearic Basin
which descends rapidly off the coasts of North Africa, the
French Riviera, Corsica and Sardinia to a maximum depth of
3,180 metres off the west coast of the latter island. Off
the coasts of eastern Spain and in the Golfe du Lion there
is a continental shelf with a maximum breadth of 60 km.; b)
the Tyrrhenian Basin which is relatively shallow in the
north between Corsica and the Italian mainland but descends
rapidly further south to a maximum depth of 3,731 metres. In
the southeast of this basin, north of Sicily, there are
volcanic islands and submarine peaks.

The eastern part can also be divided into several
distinct areas: a) the Ionian Basin, which extends from the
Strait of Otranto southward to the coasts of Libya and
Tunisia. It descends to nearly 5,000 metres south of Cape

Matapan, and has practically no islands. There is a broad
continental shelf in the Gulf of Gabes; b) the Levantine
Basin which forms the southwestern part of the Mediterranean. This basin is separated from the Ionian Basin by a
submarine ridge or sill located between Crete and Libya. It
descends to about 4,500 metres near Rhodes, and contains a
sedimentary continental shelf over 100 km. broad off the
Nile estuary. The most significant island in the area is
Cyprus; c) the Adriatic Sea which is 770 km. long and has a
maximum width of 200 km. Its average depth is about 250
metres and most of it is shallower than this, making it the
shallowest part of the Mediterranean; d) the Aegean Sea
which is an area of geographical instability and subsidence.
The peninsulas and islands thereof continue the geographical
trend and structure of the adjacent mainlands.

The submarine platform linking the two parts of the
Mediterranean is less than 200 metres deep over most of its
area. It contains several Italian islands including
Lampedusa, Pantelleria and Linosa, as well as the Maltese
islands.

An exchange of water, partly in the form of deep
currents, takes place with the Atlantic Ocean and there is a
smaller exchange with the Black Sea. The inflow of water
through runoff and precipitation is less than the loss
through evaporation, causing the salinity of the Mediterranean to be higher than in the neighbouring Atlantic.

The Mediterranean Sea is connected to the Atlantic
Ocean by the Strait of Gibraltar. It is connected to the
Black Sea by the Dardanelles, which at its narrowest is
about 1,700 metres wide, the Sea of Marmara and the
Bosphorus. The Strait of Gibraltar is about 58 km. long and
at its narrowest about 13 km. wide between Point Marroque in
Spain and Point Cires in Morocco. Both Morocco and Spain
claim a territorial sea of 12 nautical miles which means
that most of the Strait lies in the territorial sea of one
state or the other. As Spain possesses an enclave, Ceuta, on
the southern coast of the Strait and at its extreme east
end, in theory the whole width of the Strait for a few
kilometres consists of Spanish territorial sea. The bed of
the Strait is irregular with some submarine canyons, and
fast currents pass through it both at the surface and near
to the bed. The deepest part of the Strait, about 300
metres, is located near Tangier.

THE LEGAL BACKGROUND

No apology is needed for asserting the importance of legal considerations to the subject matter of this volume. For anyone with doubts, it is recommended that he read the late Professor D.P. O'Connell's The Influence of Law on Sea Power, 1975.

The International Law of the Sea: the 1958 Conventions

Before the particular problems of national sea boundaries in the Mediterranean can be discussed, it is necessary to explain briefly the general background to the international law of the sea.

Both the League of Nations and its successor the United Nations recognised the law of the sea to be a subject suitable for consolidation in a multilateral convention. The League of Nations Codification Conference in 1930 agreed on draft articles to regulate the legal regimes of internal waters and territorial seas but could not agree on a uniform breadth for the latter. The subject of the law of the sea was placed early on the agenda of the International Law Commission, a body of jurists appointed by the General Assembly to fulfil its function under Article 13(1) of the Charter of "promoting international cooperation in the political field and encouraging the progressive development of international law and its codification". After six years' work the Commission produced draft articles in 1956 which were submitted to the First United Nations Conference on the Law of the Sea in 1958. This Conference resulted in the signature, by the majority of the states then in existence, of four conventions, dealing respectively with the Territorial Sea and the Contiguous Zone, the High Seas, the Continental Shelf, and the Fishing and Conservation of the Living Resources of the High Seas. The Second United Nations Conference in 1960 narrowly failed to reach agreement on a uniform breadth for the territorial sea. After nine years of deliberations, the Third United Nations Conference on the Law of the Sea (referred to hereafter as UNCLOS III) agreed in April 1982, by 130 votes to 4 with 17 abstentions, to proceed to the signature later that year of a single convention which would cover most aspects of the subject. This convention (referred to hereafter as the 1982 Convention) was signed in December 1982.

In the following section of this chapter, the legal

concepts relevant to a discussion of national sea boundaries in the Mediterranean will be explained in some detail, taking the 1958 Conventions as the historical starting-point.

All four Conventions are in force for the states which have ratified or acceded to them but the number of states parties to them is not great in proportion to the total number of states now in existence, or even to the total number of coastal states. Thus the High Seas Convention has about 60 parties, the Territorial Sea and Contiguous Zone Convention about 50, the Continental Shelf Convention about 55 and the Fisheries Convention about 50. Three of the Conventions, Territorial Sea, Continental Shelf, and High Seas, are to a large extent codifications of existing rules of customary international law or reflect general principles of international law. Thus the fact that many states are not parties to some or any of these three Conventions does not mean that such rules and principles are not binding on those states in any form. It will be as customary rules or as general principles that they will be binding, however, not as treaty rules. The practice of the Mediterranean states in respect of the Conventions will be discussed later.

Relevant Concepts in the International Law of the Sea

Internal waters. Article 5(1) of the 1958 Convention on the Territorial Sea and Contiguous Zone reads: "Waters on the landward side of the baseline of the territorial sea form part of the internal waters of the State."

A coastal state exercises sovereignty over these internal waters, and equally over their subjacent bed and subsoil and superjacent airspace. Although this is not stated expressly in the 1958 Convention, it is clear that this is the case both on the basis of the customary international law which applied at the time of the Convention, and still applies, and as a deduction a fortiori from the Convention's treatment of the concept of the territorial sea which lies seaward of internal waters.

Included in internal waters are ports, harbours, estuaries and bays, the latter being defined in a complex provision, Article 7, which lays down a maximum closing line of 24 nautical miles. Also within the concept of internal waters are so-called "historic bays", even if in excess of 24 miles in breadth, as well as waters enclosed within a system of straight baselines parallel to the general direction of the coast which the coastal state may draw where the coastline is deeply indented and cut into, or where there is a fringe of islands along the coast. The use of the system of straight baselines to enclose such features was recognised as lawful by the International Court of Justice in the Anglo-Norwegian Fisheries Case in 1951 and is now provided for in Article 4 of the Convention of 1958.

The significance of internal waters for the purposes of this paper is that as the coast state has full sovereignty therein it has the right to forbid navigation and any other activity on, under or over them. There is some mitigation of the rigour of this conclusion in respect of sea areas which have been enclosed by straight baselines. If the areas in question were previously part of the territorial sea or high seas, a right of innocent passage for foreign ships is preserved therein. The meaning of "innocent passage" will be discussed below.

Territorial seas. Article 1(1) of the 1958 Convention reads: "The sovereignty of a State extends, beyond its land territory and its internal waters, to a belt of sea adjacent to its coast, described as the territorial sea."

Article 2 reads: "The sovereignty of a coastal State extends to the air space over the territorial sea as well as to its bed and subsoil."

The baseline for the measurement of the territorial sea is normally the low water line along the coast as marked on large-scale charts recognised by the coastal state, although the baseline may also be at the seaward limit of internal waters, e.g., the closing line in bays. Article 6 of the 1958 Convention states tht "the outer limit of the territorial sea is the line every point of which is at a distance from the nearest point of the baseline equal to the breadth of the territorial sea." The Convention, however, does not prescribe the maximum limit of the territorial sea, and, as already mentioned, the Second United Nations Conference failed to secure agreement on this in 1960.

The sovereignty of the coastal state over its territorial sea is subject to the right of innocent passage for the ships of all other states whether coastal or not. Passage is defined in Article 14(2) of the 1958 Convention to comprise navigation through the territorial sea for the purpose of traversing that sea without entering internal waters, or of proceeding to internal waters, or of making for the high seas from internal waters. Article 14(4) declares that "passage is innocent so long as it is not prejudicial to the peace, good order or security of the coastal State," and a coastal state is permitted to prevent passage which is not innocent.

Article 17 provides that foreign ships in innocent passage "shall comply with the laws and regulations enacted by the coastal State in conformity with these Articles and other rules of international law and, in particular, with such laws and regulations relating to transport and navigation." Article 19 enjoins the coastal state, however, not to exercise its criminal jurisdiction on board a foreign ship in innocent passage save only if (i) the consequences of the crime extend to the coastal state, or (ii) if the crime is of a kind to disturb the peace of the coastal state or the good order of the territorial sea, or (iii) the assistance of the local authorities has been requested by the captain or consul of the flag-state, or, (iv) such exercise of jurisdiction is necessary for the suppression of illicit traffic in narcotic drugs. Nor may a coastal state exercise civil jurisdiction on board a foreign ship in innocent passage.

The test of "innocence" is still a subject of controversy. The view favoured by ship-owning states is that it is objective, and only the manner of passage may be innocent or non-innocent, and not extraneous factors such as the cargo carried, the type of ship, or potential as opposed to existing risks. On the other hand, the view favoured by some coastal states is that these factors and objective considerations such as motive may be taken into account in assessing whether passage is innocent or not.

The provisions regarding innocent passage apply not only to privately-owned merchant ships but also to government-owned commercial vessels as well as to warships. In the case of warships, there is a long-standing controversy over whether they have a right to enter the territorial sea without the prior permission of the coastal state or at least notification thereto. This is not expressly clarified in the 1958 Convention although by Article 23 a coastal state is empowered to require a foreign warship to leave the territorial sea if it does not comply with the local regulations for passage and disregards any request for compliance with them. Foreign submarines are obliged to navigate on the surface and to show their flag.

Article 16(3) of the 1958 Convention permits the coastal state, without discriminating among foreign ships, to suspend temporarily by notice in specified areas of its territorial sea the innocent passage of foreign ships if such suspension "is essential for the protection of its security". As will be seen later, the power to suspend innocent passage does not apply where the territorial sea is part of a strait used for international navigation.

The delimitation of the territorial sea is provided for in Article 12(1) of the 1958 Convention as follows:

> Where the coasts of two States are opposite or adjacent to each other, neither of the two States is entitled, failing agreement between them to the contrary, to extend its territorial sea beyond the median line every point of which is equidistant from the nearest points on the baselines from which the breadth of the territorial seas of each of the two States is measured. The provisions of this paragraph shall not apply, however, where it is necessary by reason of historic title or other special circumstances to delimit the territorial seas of the two States in a way which is at variance with this provision.

Contiguous zones. Article 24 provides as follows:

> 1. In a zone of the high seas contiguous to its territorial sea, the coastal State may exercise the control necessary to:
>   (a) Prevent infringement of its customs, fiscal, immigration or sanitary regulations within its territory or territorial sea;
>   (b) Punish infringement of the above regulations committed within its territory or territorial sea.
> 2. The contiguous zone may not extend beyond twelve miles from the baseline from which the breadth of the territorial sea is measured.
> 3. Where the coasts of two States are opposite or adjacent to each other, neither of the two States is entitled, failing agreement between them to the contrary, to extend its contiguous zone beyond the median line every point of which is equidistant from the nearest points on the base-

lines from which the breadth of the territorial
seas of the two States is measured.

As most of the Mediterranean states now have a twelve-
mile territorial sea this provision of the 1958 Convention
has lost much of its relevance. As will be shown later, the
concept of the contiguous zone remains in the 1982 Convention where it is extended to a maximum distance of 24 miles.

Straits. Article 16(4) provides that there shall be no
suspension of innocent passage "through straits which are
used for international navigation between one part of the
high seas and another part of the high seas or the territorial sea of a foreign State". There is a strong inference,
from the fact that the provision is placed in the section of
the Convention which deals with innocent passage through the
territorial sea, that the other provisions relating to innocent passage, including the requirement that submarines
navigate on the surface, apply in those straits which are
made up entirely of the territorial seas of the coastal
states.

This raises again the question whether a foreign warship has a right of innocent passage through a strait
without the prior authorisation of the coastal state or
states. The International Court of Justice in the Corfu
Channel Case in 1949 considered that such a right existed
under customary law. It also rejected an argument by Albania
that in order to classify as a strait the feature had to be
essential to passage between the particular two sections of
the high seas in question.

Continental shelves. Article 1 of the 1958 Convention
on the Continental Shelf defines the concept as follows:

> For the purpose of these Articles, the term 'con-
> tinental shelf' is used as referring (a) to the
> seabed and subsoil of the submarine areas adjacent
> to the coast but outside the areas of the terri-
> torial sea, to a depth of 200 metres or, beyond
> that limit, to where the depth of the superjacent
> waters admits of the exploitation of the natural
> resources of the said areas; (b) to the seabed and
> subsoil of similar submarine areas adjacent to the
> coasts of islands.

Article 2 reads:

1. The coastal state exercises over the continental shelf sovereign rights for the purpose of exploring it and exploiting its natural resources.
2. The rights referred to in paragraph 1 of this article are exclusive in the sense that if the coastal State does not explore the continental shelf or exploit its natural resources, no one may undertake these activities, or make a claim to the continental shelf, without the express consent of the coastal State.
3. The rights of the coastal State over the continental shelf do not depend on occupation, effective or notional, or on any express proclamation.
4. The natural resources referred to in these Articles consist of the mineral and other non-living resources of the seabed and subsoil together with living organisms belonging to sedentary species, that is to say, organisms which, at the harvestable stage, either are immobile on or under the seabed or are unable to move except in constant physical contact with the seabed or the subsoil.

Article 3 provides:

The rights of the coastal State over the continental shelf do not affect the legal status of the superjacent waters as high seas, or that of the air space above those waters.

These three Articles were considered by the International Court of Justice in the North Sea Continental Shelf Cases in 1969 to reflect existing customary international law and thus applicable even to states which were not parties to the Convention.

Article 6 of the Continental Shelf Convention sets out the rules for the delimitation of opposite and adjacent states. The respective provisions run as follows:

1. Where the same continental shelf is adjacent to

the territories of two or more States whose coasts are opposite each other, the boundary of the continental shelf appertaining to such States shall be determined by agreement between them. In the absence of agreement and unless another boundary line is justified by special circumstances, the boundary is the median line, every point of which is equidistant from the nearest points of the baselines from which the breadth of the territorial sea of each State is measured.

2. Where the same continental shelf is adjacent to the territorial of two adjacent States, the boundary of the continental shelf shall be determined by agreement between them. In the absence of agreement, and unless another boundary line is justified by special circumstances, the boundary shall be determined by application of the principle of equidistance from the nearest points of the baselines from which the breadth of the territorial sea of each State is measured.

In the North Sea Continental Shelf Cases the International Court of Justice considered that Article 6(2) did not reflect an existing customary rule in being at the time the Convention was concluded, nor had it become such a rule by subsequent state practice. Consequently, it was not binding on the Federal Republic of Germany, a non-party to the Convention. The same reasoning probably applies to Article 6(1).

High seas. The 1958 Geneva Convention on the High Seas provides for the legal status of "all parts of the sea that are not included in the territorial sea or in the internal waters of a State". That this definition includes the waters superjacent to the continental shelf is expressly confirmed by Article 3 of the Continental Shelf Convention. It is controversial whether it applies to the bed and subsoil of the sea beyond the continental shelf.

The status of the high seas as so defined is set out in Article 2 of the Convention on the High Seas as follows:

The high seas being open to all nations, no State may validly purport to subject any part of them to its sovereignty. Freedom of the high seas is exercised under the conditions laid down by these articles and by the other rules of international law. It comprises, _inter alia_, both for coastal and non-coastal States: 1) Freedom of navigation; 2) Freedom of fishing; 3) Freedom to lay submarine cables and pipelines; 4) Freedom to fly over the high seas. These freedoms, and others which are recognised by the general principles of international law, shall be exercised by all States with reasonable regard to the interests of other States in their exercise of the freedom of the high seas.

Article 6(1) confirms a basic rule of customary international law: "Ships shall sail under the flag of one State only, and, save in exceptional cases expressly provided for in international treaties or in these Articles, shall be subject to its exclusive jurisdiction on the high seas."

Two such "exceptional cases" provided for elsewhere in the Convention are piracy and slave trade. Articles 8(1) and 9 provide that warships and state-owned or operated ships used on government non-commercial service shall be immune on the high seas from the jurisdiction of any other state.

The 1958 Convention on Fishing and Conservation of the Living Resources of the High Seas provides in Article 1(2) that "all States have the duty to adopt, or to co-operate with other States in adopting, such measures for their respective nationals, as may be necessary for the conservation of the living resources of the high seas". Only three Mediterranean states are parties to this Convention, however, and the development of the concept of the exclusive fishing zone lying outside the territorial sea has diminished its significance.

Hot pursuit. Article 23 of the 1958 High Seas Convention permits a coastal state to pursue and arrest a foreign ship on the high seas if the authorities of the coastal state "have good reason to believe that the ship has violated the laws and regulations of that State". The pursuit must start when the foreign vessel is within the internal waters, territorial sea or contiguous zone of the pursuing state. If within the contiguous zone, pursuit may only be undertaken for violation of the rights for the protection of which the zone was established.

Islands. Article 10(1) of the Convention on the Territorial Sea and the Contiguous Zone defines an island as "a naturally-formed area of land, surrounded by water, which is above water at high tide". Article 10(2) then provides that "the territorial sea of an island is measured in accordance with the provisions of these Articles" which means, for example, that the normal baseline of an island is the low-water line along the coast. An island as defined above is treated, in other words, just like any other part of state territory. Article 1 of the Convention on the Continental Shelf applies the term "continental shelf" equally to submarine areas adjacent to the coasts of islands.

The International Law of the Sea: The Work of UNCLOS III

The work of the United Nations which resulted in December 1982 in the signature of the 1982 Convention on the Law of the Sea can be said to have started in 1967 when Mr. Pardo, Ambassador of Malta, introduced in the First Committee of the General Assembly of the United Nations an agenda item relating to the peaceful use of the seabed and ocean floor "beyond the limits of present nationaljurisdiction". This led to the establishment of an ad hoc Committee to study the subject, the passing by the General Assembly on 17 December 1970 of the "Declaration of Principles", and the decision of the General Assembly to convene a new conference to consider the law of the sea in general. The Conference met in New York in December 1973 and regularly thereafter. Over 150 States were represented. The Conference considered a number of texts headed successively Main Trends, Informal Single Negotiating Text, Revised Single Negotiating Text, Informal Composite Negotiating Text, Draft Convention (Informal Text) and Draft Convention. Some of the provisions underwent little or no change from one text to another, reflecting the view that the law therein was regarded as already substantially settled; other provisions were subject to drastic changes in the course of the life of the Conference.

The concepts already discussed in the context of the 1958 Conventions will be re-examined here in the light of the text of the 1982 Convention.

Internal waters. The provisions of the 1982 Convention are substantially similar to those in the 1958 Convention on the Territorial Sea and the Contiguous Zone.

Territorial seas. Article 3 of the 1982 Convention has made a substantial addition to that appearing in the 1958 Convention; it reads: "Every State has the right to establish the breadth of its territorial sea up to a limit not exceeding 12 nautical miles, measured from baselines determined in accordance with this Convention."

Article 15 relating to delimitation of territorial seas between States with opposite coasts is in terms similar to Article 12 of the 1958 Convention discussed above.

The concept of innocent passage for foreign ships through the territorial sea is defined in the same terms as in Article 14(4) of the 1958 Convention. The 1982 Convention, however, enumerates in Article 19 eleven activities which would render the passage of the foreign ship prejudicial to the peace, good order or security of the coastal state and thus "non-innocent". These are: a) any threat or use of force against the sovereignty, territorial integrity or political independence of the coastal state, or in any other manner in violation of the principles of international law embodied in the Charter of the United Nations; b) any exercise or practice with weapons of any kind; c) any act aimed at collecting information to the prejudice of the defence or security of the coastal state; d) any act of propaganda aimed at affecting the defence or security of the coastal state; e) the launching, landing or taking on board of any aircraft; f) the launching, landing or taking on board of any military devices; g) the loading or unloading of any commodity, currency or person contrary to the customs, fiscal, immigration or sanitary laws and regulations of the coastal state; h) any act of wilful and serious pollution contrary to this Convention; i) any fishing activities; j) the carrying out of research or survey activities; k) any act aimed at interfering with any systems of communication or any other facilities or installations of the coastal state. A final residual item covers "any other activity not having a direct bearing on passage".

The 1982 Convention goes on to provide in Article 21 that the coastal state may adopt laws and regulations relating to innocent passage in respect of all or any of the following: a) the safety of navigation and the regulation of maritime traffic; b) the protection of navigational aids and facilities and other facilities or installations; c) the protection of cables and pipelines; d) the conservation of the living resources of the sea; e) the prevention of infringement of the fisheries laws and regulations of the coastal state; f) the preservation of the environment of the coastal state and the prevention, reduction and control of pollution thereof; g) marine scientific research and hydrographic surveys; h) the prevention of infringement of the customs, fiscal, immigration or sanitary laws and regulations of the coastal state.

Article 22 permits the coastal state to establish sea lanes and traffic separation schemes in the territorial sea.

Article 24, however, limits the power of the coastal state to control innocent passage in its territorial sea in the following respects:

Duties of the coastal State
1. The coastal State shall not hamper the innocent passage of foreign ships through the territorial sea except in accordance with this Convention. In particular, in the application of this Convention or of any laws or regulations adopted in conformity with this Convention, the coastal State shall not:
    (a) impose requirements on foreign ships which have the practical effect of denying or impairing the right of innocent passage, or
    (b) discriminate in form or in fact against the ships of any State or against ships carrying cargoes to, from or on behalf of any State.
2. The coastal State shall give appropriate publicity to any danger to navigation, of which it has knowledge, within its territorial sea.
    q

Article 25 provides in terms similar to Article 16(1) and (3) of the 1958 Convention that:

1. The coastal State may take the necessary steps in its territorial sea to prevent passage which is not innocent.
2. The coastal State may, without discrimination in form or in fact among foreign ships, suspend temporarily in specified areas of its territorial sea the innocent passage of for-

eign ships if such suspension is essential for the protection of its security, including weapons exercises. Such suspicions shall take effect only after having been duly published.

The 1982 Convention has not clarified the doubt over whether a warship needs prior permission of, or must make prior notification to the coastal state for passage through its territorial sea. Article 30 of the 1982 Convention follows Article 23 of the 1958 Convention in requiring warships to comply with the laws and regulations of the coastal state concerning passage and in empowering the coastal state to order the warship to leave the territorial sea if it fails to comply. In other respects, however, warships remain immune from the enforcement jurisdiction of the coastal state.

Contiguous zones. Article 33 of the 1982 Convention repeats substantially the provisions of Article 24 of the 1958 Convention, with the important alteration that the minimum distance of the zone is increased from 12 to 24 nautical miles from the baseline of the territorial sea.

Straits. The 1982 Convention makes considerable changes to the 1958 regime and it might be said this is one of the Convention's major features. A new concept of "transit passage" has been devised which by Article 44 is not suspendable by the coastal state and so differs from the concept of "innocent passage" in the 1958 Convention. It also differs in quality from "non-suspendable innocent passage" in Article 16(4) of the 1958 Convention with fewer restrictions on the vessel in transit. Furthermore, it applies to aircraft as well as ships.

Article 37 of the 1982 Convention applies transit passage to: "... straits which are used for international navigation between one part of the high seas or an exclusive economic zone and another part of the high seas or an exclusive economic zone."

It would appear that such straits must fall within the territorial sea of one or the other or both of the coastal states since there is nothing in the 1982 Convention (or in the 1958 Convention) to permit the coastal states to control a strait beyond the extent of their respective territorial seas simply on the ground that it is a geographical strait.

Not all straits, however, fall within the scope of transit passage. Article 38(1) of the 1982 Convention excludes transit passage from straits "formed by an island of a State bordering the strait and its mainland ... if there exists seaward of the island a route through the high seas or through an exclusive economic zone of similar convenience with respect to navigational and hydrographical character- istics." In such straits a concept of non-suspendable inno- cent passage is applied by Article 45.

Article 38(2) of the 1982 Convention defines transit passage as: "... the exercise ... of the freedom of naviga- tion and overflight solely for the purpose of continuous and expeditious transit of the strait between one part of the high seas or an exclusive economic zone and another part of the high seas or an exclusive economic zone."

The duties of ships and aircraft during transit pas- sage are set out in Articles 38 and 40 of the 1982 Conven- tion. In particular, they must refrain from any political threat or use of force against the sovereignty, territorial integrity or political independence of states bordering the strait and must comply with "generally accepted" inter- national regulations, procedures and practices for safety at sea and for the prevention, reduction and control of pollu- tion from ships. Research and survey activities can only be carried out with the permission of the coastal states.

Article 42 sets out the extent to which the states bordering straits may adopt laws and regulations relating to transit passage. Such laws and regulations, which must not discriminate among foreign ships and which must not have the practical effect of denying, hampering or impairing transit passage, are confined to all or any of the following:

a) the safety of navigation and the regulation of maritime traffic, as provided in article 41; (i.e. sea lanes and traffic separation schemes);

b) the prevention, reduction and control of pollu- tion, by giving effect to applicable inter- national regulations regarding the discharge of oil, oily wastes and other noxious sub- stances in the strait;

c) with respect to fishing vessels, the preven- tion of fishing, including the stowage of fish- ing gear;

d) the loading or unloading of any commodity, currency or person in contravention of the

customs, fiscal, immigration or sanitary laws
and regulations of States bordering straits.

Article 42(4) provides that foreign ships exercising
the right of transit passage shall comply with such laws and
regulations.

The enforcement jurisdiction of the coastal state, as
distinguished from its prescriptive jurisdiction, over
foreign vessels passing through straits is constrained by
Article 233 which in effect permits the coastal state to
take "appropriate enforcement measures" only where the vessel has committed a violation of the laws and regulations
referred to in Article 42, paragraph 1(a) and (b), "causing
or threatening major damages to the marine environment of
the straits."

The Straits of Gibraltar would appear to be subject to
the regime of transit passage provided in the Draft Convention. The Dardanelles, on the other hand, would be excluded from the operation of the 1982 Convention since by
Article 35(c) the Convention does not affect "the legal
regime in straits in which passage is regulated in whole or
in part by long-standing international conventions in force
specifically relating to such straits". Passage through
the Dardanelles, as well as through the Sea of Marmara and
the Bosphorus, is regulated by the Montreux Convention of
1936.

Continental shelves. The 1982 Convention has brought
about signficant changes in the extent and delimitation of
the continental shelf. Article 76 reads:

1. The continental shelf of a coastal State comprises the sea-bed and subsoil of the submarine areas that extend beyond its territorial sea throughout the natural prolongation
   of its land territory to the outer edge of the
   continental margin, or to a distance of 200
   nautical miles from the baselines from which
   the breadth of the territorial sea is measured
   where the outer edge of the continental margin
   does not extend up to that distance.
2. The continental shelf of a coastal State shall
   not extend beyond the limits provided for in
   paragraph 4 to 6.
3. The continental margin comprises the submerged

prolongation of the land mass of the coastal State, and consists of the sea-bed and subsoil of the shelf, the slope and the rise. It does not include the deep ocean floor with its oceanic ridges or the subsoil thereof.

4. (a) For the purposes of this Convention, the coastal State shall establish the outer edge of the continental margin wherever the margin extends beyond 200 nautical miles from the baselines from which the breadth of the territorial sea is measured, by either:

    (i) a line delineated in accordance with paragraph 7 by reference to the outermost fixed points at each of which the thickness of sedimentary rocks is at least 1 per cent of the shortest distance from such point to the foot of the continental slope; or

    (ii) a line delineated in accordance with paragraph 7 by reference to fixed points not more than 60 nautical miles from the foot of the continental slope.

(b) In the absence of evidence to the contrary, the foot of the continental slope shall be determined as the point of maximum change in the gradient at its base.

5. The fixed points comprising the line of the outer limits of the continental shelf on the sea-bed, drawn in accordance with paragraph 4(a) (i) and (ii), either shall not exceed 350 nautical miles from the baselines from which the breadth of the territorial sea is measured or shall not exceed 100 nautical miles from the 2,500 metre isobath, which is a line connecting the depth of 2,500 metres.

6. Notwithstanding the provisions of paragraph 5, on submarine ridges, the outer limit of the continental shelf shall not exceed 350 nautical miles from the baselines from which the breadth of the territorial sea is measured. This paragraph does not apply to submarine elevations that are natural components of the

> continental margin, such as its plateaux,
> rises, caps, banks and spurs.
>
> 7. The coastal State shall delineate the outer
>   limits of its continental shelf, where that
>   shelf extends beyond 200 nautical miles from
>   the baselines from which the breadth of the
>   territorial sea is measured, by straight lines
>   not exceeding 60 nautical miles in length,
>   connecting fixed points, defined by co-
>   ordinates of latitude and longitude.

The effect of this provision is to separate the legal
concept of the shelf from the exploitability criterion set
out in the 1958 Continental Shelf Convention and indeed to
separate it from the physical presence of a continental
margin since a state without a continental margin in the
physical sense is nevertheless entitled to a continental
shelf in the juridical sense to a distance of 200 nautical
miles.

Article 83(1) of the 1982 Convention has substantially
altered the criterion for delimitation of the shelf between
States with opposite or adjacent coasts. It reads:

> The delimitation of the continental shelf bet-
> ween States with opposite or adjacent coasts
> shall be effected by agreement on the basis of
> international law, as referred to in Article 38
> of the Statute of the International Court of
> Justice, in order to achieve an equitable solu-
> tion.

Paragraphs 2 and 3 of the Article read:

> 2. If no agreement can be reached within a reason-
>   able period of time, the States concerned
>   shall resort to the procedures provided for in
>   Part XV (i.e., the settlement of disputes
>   provision);
> 3. Pending agreement as provided for in paragraph
>   1, the States concerned, in a spirit of under-
>   standing and co-operation, shall make every
>   effort to enter into provisional arrangements
>   of a practicl nature and, during this tran-
>   sitional period, not to jeopardize or hamper

the reaching of the final agreement. Such
arrangements shall be without prejudice to the
final delimitation.

High seas. The concept of high seas has been affected
in an important respect by the concept of the exclusive
economic zone found in the 1982 Convention. This latter
concept will be dealt with fully in the next section of this
paper.

Having defined high seas in Article 86 as "all parts
of the sea that are not included in the exclusive economic
zone, in the territorial sea or in the internal waters of a
State, or in the archipelagic waters of an archipelagic
State", the 1982 Convention in Article 87 provides:

1. The high seas are open to all States, whether
   coastal or landlocked. Freedom of the high
   seas is exercised under the conditions laid
   down by this Convention and by other rules of
   international law. It comprises, inter alia,
   both for coastal and landlocked States: (a)
   freedom of navigation; (b) freedom of overflight; (c) freedom to lay submarine cables
   and pipelines, subject to Part VI; (d) freedom
   to construct artificial islands and other installations permitted under international law,
   subject to Part VI; (e) freedom of fishing,
   subject to the conditions laid down in section
   2; (f) freedom of scientific research, subject
   to Parts VI and XIII.
2. These freedoms shall be exercised by all
   States with due regard for the interests of
   other States in their exercise of the freedom
   of the high seas, and also with due regard for
   the rights under this Convention with respect
   to activities in the Area.

Article 88 provides that the high seas shall be re-
served for peaceful purposes while Article 89 proclaims that
"no State may validly purport to subject any part of the
high seas to its sovereignty". Article 92 repeats Article 6
of the 1958 Convention in declaring that "ships shall sail
under the flag of one State only and, save in exceptional
cases expressly provided for in international treaties or in
94

this Convention, shall be subject to its exclusive jurisdiction on the high seas." Piracy and the slave trade are again expressly provided to be exceptions.

Article 95 and 96 provide for the complete immunity of warships and state-owned or operated ships on governmental non-commercial service on the high seas from the jurisdiction of any state other than the flag state.

Hot pursuit. The 1982 Convention has made a substantial broadening of the range of hot pursuit in order to accommodate the doctrine to the new maritime zones over which the coastal state has jurisdiction and which will be discussed in the next section of this paper. Article 111 of the text reads:

1. The hot pursuit of a foreign ship may be undertaken when the competent authorities of the coastal State have good reason to believe that the ship has violated the laws and regulations of that State. Such pursuit must be commenced when the foreign ship or one of its boats is within the internal waters, the archipelagic waters, the territorial sea or the contiguous zone of the pursuing State, and may only be continued outside the territorial sea or the contiguous zone if the pursuit has not been interrupted. It is not necessary that, at the time when the foreign ship within the territorial sea or the contiguous zone receives the order to stop, the ship giving the order should likewise be within the territorial sea or the contiguous zone. If the foreign ship is within a contiguous zone, as defined in Article 33, the pursuit may only be undertaken if there has been a violation of the rights for the protection of which the zone was established.

2. The right of hot pursuit shall apply _mutatis mutandis_ to violations in the exclusive economic zone or on the continental shelf, including safety zones around continental shelf installations, of the laws and regulations of the coastal State applicable in accordance with this Convention to the exclusive economic zone or the continental shelf, including such

safety zones.

Islands. In the 1982 Convention there is a separate Article 121 which relates to islands. This makes a drastic change by reason of the effect of paragraph 3. The Article reads:

1. An island is a naturally formed area of land, surrounded by water, which is above high water at high tide.
2. Except as provided for in paragraph 3, the territorial sea, the contiguous zone, the exclusive economic zone and the continental shelf of an island are determined in accordance with the provisions of this Convention applicable to other land territory.
3. Rocks which cannot sustain human habitation or economic life of their own shall have no exclusive economic zone or continental shelf.

Thus, certain islands are not capable of generating rights beyond 12 miles from their baseline.

The International Law of the Sea: The Work of UNCLOS III in Developing New Concepts

The 1982 Convention has introduced certain important concepts not found expressly in the 1958 Conventions. One of these, transit passage through straits, has been discussed above. This section is dedicated to the examination of the others.

Archipelagic states. The 1982 Convention defines an "archipelagic State" as one constituted wholly by one or more archipelagos and may include other islands. The term "archipelago" means a group of islands, including parts of islands, interconnecting waters and other natural features which are so closely interrelated that such islands, waters and other natural features form an intrinsic geographical, economic and political entity, or which historically have been regarded as such.

The 1982 Convention permits such a state to draw straight baselines joining the outermost points of the islands within certain specified tolerances. Within the baselines, other than in areas of internal waters, foreign ships enjoy a right of innocent passage, suspendable on a non-discriminatory basis if essential for the protection of the security of the archipelagic state. The archipelagic state may designate sea lanes and air routes suitable for the continuous and expeditious passage of foreign ships and aircraft through or over its archipelagic waters. All ships and aircraft enjoy the right of archipelagic sea lanes passage in such sea lanes and air routes.

In the Mediterranean, only Malta would seem to qualify as an archipelagic state under the above definition. The Maltese islands are compact and the total area of sea classified as archipelagic waters is likely to be relatively restricted.

Exclusive economic zones. In the years following the conclusion of the 1958 Conventions one of the most significant developments in the international law of the sea was the emergence of the concept of an exclusive economic zone to embrace not only the natural resources of the continental shelf but also of the water superjacent thereto. It thus extended beyond the seaward limits of the territorial sea and into the area defined as "high seas" in the 1958 High Seas Convention. The exclusive economic zone had its origins in part in the practice of states after 1958 in concluding bilateral agreements establishing exclusive fisheries zones. The European Fisheries Convention of 1964, concluded between twelve European States, including three Mediterranean states and the United Kingdom, provided for an exclusive fisheries zone for each party measured 12 miles from the baseline of the territorial sea. Thereafter the breadth of national exclusive fisheries zones, particularly in South America and Africa, widened as far as 200 miles. In the Anglo-Icelandic Fisheries Case in 1974, the International Court of Justice, though holding that a 50-mile Icelandic exclusive fisheries zone was not opposable to the United Kingdom, refrained from pronouncing it invalid erga omnes. State practice continued to regard such zones as lawful, and in 1977 some member states of the European Economic Community proclaimed national exclusive fishing zones of 200 miles in the North Sea and the Atlantic.

Meanwhile, state practice, particularly in South America, was establishing the legality of a zone, called variously the patrimonial sea, the epicontinental sea or the exclusive economic zone, which extended to a maximum distance of 200 nautical miles from the baseline of the territorial sea. Article 1 of the Declaration of Santa Domingo in June

1972 illustrated the concept. It stated: "The coastal State has sovereign rights over the renewable and non-renewable natural resources which are found in the waters, in the seabed and in the subsoil of an area adjacent to the territorial sea called the patrimonial sea." The European states, however, and in particular the Mediterranean states, did not at this stage assert economic zones in this sense.

The exclusive economic zone is defined in Articles 55 and 57 of the 1982 Convention as an area beyond and adjacent to the territorial sea not extending beyond 200 nautical miles from the baselines from which the breadth of the territorial sea is measured. Article 56 sets out the rights, jurisdiction and duties of the coastal state in the zone as follows:

> (a) sovereign rights for the purpose of exploring and exploiting, conserving and managing the natural resources, whether living or non-living, of the waters superjacent to the sea bed and of the seabed and its subsoil, and with regard to other activities for the economic exploitation and exploration of the zone, such as the production of energy from the water, currents and winds;
> (b) jurisdiction as provided for in the relevant provisions of this Convention with regard to: (i) the establishment and use of artificial islands, installations and structures; (ii) marine scientific research; (iii) the protection and preservation of the marine environment;
> (c) other rights and duties provided for in this Convention.

By Article 60 the coastal state is also given the exclusive right therein to construct and regulate artificial islands, installations and structures "for the purposes provided for in Article 56 and other economic purposes".

A key provision is Article 58 which provides:

> In the exclusive economic zone, all States, whether coastal or land-locked, enjoy, subject to the relevant provisions of this Convention, the freedoms ... of navigation and overflight and of

the laying of submarine cables and pipelines, and other internationally lawful uses of the sea related to these freedoms, such as those associated with the operation of ships, aircraft and submarine cables and pipelines, and compatible with other provisions of this Convention.

It seems from this provision that the exclusive economic zone is to be equated with the high seas rather than with an area sui generis. The opinion that it is equated with high seas is strengthened by Article 58(2) of the 1982 Convention which reads: "Articles 88 to 115 and other pertinent rules of international law apply to the exclusive economic zone in so far as they are not incompatible with this Part."

These Articles are located in the Part of the 1982 Convention which deals with "high seas".

On the other hand, Article 86 provides that the provisions of the Part of the 1982 Convention which deals with high seas "apply to all parts of the sea that are not included in the exclusive economic zone, in the territorial sea or in the internal waters of a State, or in the archipelagic waters of an archipelagic State". Thus when Article 116 declares that "all States have the right for their nationals to engage in fishing on the high seas" the term "high seas" excludes the exclusive economic zone. By Article 61, the coastal state shall determine the allowable catch of the living resources in its exclusive economic zone. Article 62 provides that where the coastal state does not have the capacity to harvest the entire allowable catch it shall by agreement or other arrangements give other States access to the surplus of the allowable catch, having particular regard to the position of (i) land-locked States, (ii) those States whose geographical situation makes them dependent upon the exploitation of the fishery resources of the exclusive economic zones of other States, (iii) coastal States which cannot claim any exclusive economic zone of their own.

There is no freedom of scientific research in the exclusive economic zone. Article 246(1) of the 1982 Convention provides that "marine scientific research in the exclusive economic zone and on the continental shelf shall be conducted with the consent of the coastal State".

Article 74(1) provides for the delimitation of the zone in terms similar to Article 83(1) in respect of the

continental shelf: "The delimitation of the exclusive economic zone between States with opposite or adjacent coasts
shall be effected by agreement on the basis of international
law, as referred to in Article 38 of the Statute of the
International Court of Justice, in order to achieve an
equitable solution." Paragraphs 2 and 3 are similar to
paragraphs 2 and 3 of Article 83.

Among the Mediterranean states expressing views in
UNCLOS III debates, most regarded the zone as being one of
limited coastal state rights. Three such states, however,
Algeria, Libya and Albania, took a "territorial" view of the
zone. This division of opinion may become active again in
the future, particularly when states are faced with the
decision of ratifying the new Convention.

Anti-pollution measures. The 1982 Convention sets out
drastic anti-pollution measures designed to increase the
jurisdiction of the coastal states (as well as the flag
state and the port state) over pollution of the marine environment from vessels and other sources. The exclusive
economic zone is an important factor in these provisions
which increase the prescriptive as well as the enforcement
powers of the coastal state over foreign vessels in its
territorial sea and exclusive economic zone.

No attempt will be made in this paper to summarize the
relevant provisions; they are set out in Section 5 of Part
XII of the 1982 Convention, Articles 207-233.

The 1982 Convention, in Article 221, reflects the contents of the 1969 Convention relating to Intervention on the
High Seas in Cases of Oil Pollution Casualties. The Article
reads:

> Measures to avoid pollution arising from   maritime
> casualties.
> 1. Nothing in this Part shall prejudice the right
>   of States, pursuant to international law, both
>   customary and conventional, to take and en-
>   force measures beyond the territorial sea
>   proportionate to the actual or threatened
>   damage to protect their coastline or related
>   interests, including fishing, from pollution
>   or threat of pollution following upon a mari-
>   time casualty or acts relating to such a
>   casualty which may reasonably be expected to
>   result in major harmful consequence.

2. For the purpose of this article, "maritime casualty" means a collision of vessels, stranding or other incident of navigation, or other occurrence on board a vessel or external to it resulting in material damage or imminent threat of material damage to a vessel or cargo.

Enclosed or semi-enclosed seas. Part IX of the 1982 Convention consists of two Articles, 122 and 123, which have no counterpart in the 1958 Conventions. These Articles read:

Article 122 - Definition. For the purpose of this Convention 'enclosed or semi-enclosed sea' means a gulf, basin or sea surrounded by two or more States and connected to another sea or the ocean by a narrow outlet or consisting entirely or primarily of the territorial seas and exclusive economic zones of two or more coastal States.

Article 123 - Co-operation of States bordering enclosed or semi-enclosed seas. States bordering an enclosed or semi-enclosed sea should co-operate with each other in the exercise of their rights and in the performance of their duties under this Convention. To this end they shall endeavour, directly or through an appropriate regional organization:
(a) to co-ordinate the management, conservation, exploration and exploitation of the living resources of the sea;
(b) to co-ordinate the implementation of their rights and duties with respect to the protection and preservation of the marine environment;
(c) to co-ordinate their scientific research policies and undertake where appropriate joint programmes of scientific research in the area;
(d) of scientific research in the area; to invite, as appropriate, other interested States of international organisations to co-operate with them in furtherance of the provisions of this article.

In a discussion in the Second Committee of UNCLOS III on a proposal to incorporate a provision on "semi-enclosed areas" into the text, the Mediterranean states spoke with divided voices. Turkey considered that the concept of the exclusive economic zone should not be applied to the Mediterranean because, if it were, the entire sea would be subject to coastal state jurisdiction, a fact which could threaten the freedom of navigation.

Israel and Algeria both spoke in favour of the proposal. Israel considered that the freedom of navigation and overflight must be given priority in a semi-enclosed sea and that a semi-enclosed sea poor in resources such as the Mediterranean did not lend itself to far-reaching national claims.

Greece and France spoke against the proposal. Greece considered that almost all semi-enclosed seas would be covered by the general provisions in the draft articles under discussion and that existing treaties and regional agreements provided for the necessary regional cooperation to deal with pollution problems. France acknowledged that a 200-mile exclusive economic zone would place all natural resources of such seas under the coastal states' jurisdiction and asserted that it was unnecessary to demand special provisions for semi-enclosed seas in a general convention, since regional agreements already provided for in the draft text would suffice.

The Legal Status of the 1982 Convention on the Law of the Sea

The 1982 Convention is not yet a treaty document. Indeed, even after signature, it will not become a treaty document binding on those states which have signed, since by Articles 306 and 308 the Convention is subject to ratification and shall not enter into force until the lapse of 12 months from the date of deposit of the 60th instrument of ratification or accession. Thereafter it will be in force for those states which have ratified or acceded but not for other states. Taking the practice in respect of other multilateral conventions as a guide, it is likely to be some years before the document achieves the status of a treaty text.

To some extent the 1982 Convention, like the 1958 Conventions, reflects existing rules of customary international law and to this extent some of its provisions will be, and indeed already are, binding on all states as customary rules.

In its judgement in the <u>Tunisia/Libya Continental Shelf Case</u> in February 1982, the International Court of Justice remarked (paragraph 1): "(the Court) could not ignore any provision of the draft convention if it came to the conclusion that the content of such provision is binding upon all members of the international community because it embodies or crystallises a pre-existing or emergent rule of customary law."

But this cannot be said of some of the provisions which are of particular significance in the present study, for example the provisions relating to transit passage through straits, Articles 74(1) and 83(1) dealing with the delimitation of the exclusive economic zone and continental shelf respectively, and Article 76 dealing with the seaward extent of the continental shelf.

The status of the provisions regarding the exclusive economic zone is not beyond argument, although there is probably enough state practice to regard the zone as now established as a lawful extension of a coastal state's jurisdiction. What is particularly obscure in the 1982 Convention is whether the exclusive economic zone appertains to a coastal state <u>ipso jure</u> or whether its appurtenance to the coastal state arises only on the state declaring such a zone to exist. Although there is no provision dealing with the exclusive economic zone similar to Article 77(3), which provides that the rights of the coastal state over the continental shelf do not depend on occupation, effective or notional, or on any express proclamation, it would be prudent to assume that the right of a coastal state to declare such a zone creates for it some kind of inchoate right even if it has not yet made such a declaration.

In the 1977 Arbitration between France and the United Kingdom over the delimitation of the continental shelf of the Channel and its western approaches, the French Government argued that "all the Geneva Conventions on the law of the sea, including the Continental Shelf Convention, have been rendered obsolete by the recent evolution of customary law stimulated by the work of the Third United Nations Conference on the Law of the Sea." In rejecting this argument, which was opposed by the United Kingdom, the Court of Arbitration stated in its decision (paragraph 47):

... the Court recognises both the importance of

the evolution of the law of the sea which is now
in progress and the possibility that a development in customary law may, under certain conditions, ditions evidence the assent of the States
concerned to the modification, or even termina-
tion, of previously existing treaty rights and
obligations. But the Continental Shelf Convention
of 1958 entered into force as between the Parties
little more than a decade ago. Moreover, the information before the Court contains references by the
French Republic and the United Kingdom, as well as
by other States, to the Convention as an existing
treaty in force which are of quite recent date.
Consequently, only the most conclusive indications of the intention of the parties to the 1958
Convention to regard it as terminated could
warrant this Court in treating it as obsolete and
inapplicable as between the French Republic and
the United Kingdom in the present matter. In the
opinion of the Court, however, neither the records
of the Third United Nations Conference on the Law
of the Sea nor the practice of States outside the
Conference provide any such conclusive indication
that the Continental Shelf Convention of 1958 is
today considered by its parties to be already
obsolete and no longer applicable as a treaty in
force.

Despite the passage of a further five years, it is
submitted that this assessment is still correct. Although
customary international law may develop and even change with
increasing speed, there does not appear to be sufficient
state practice, apart from manifestations of support at
UNCLOS III, to crystallise as lex lata those parts of the
1982 Convention which make far-reaching changes to the 1958
regimes. The ratification or non-ratification by the individual Mediterranean states will be crucial.

The Practice of the Mediterranean States
   In respect of the 1958 Conventions. Of the nineteen
Mediterranean States (including therein the United Kingdom),
six (Spain, Italy, Malta, Israel, Yugoslavia and the United
Kingdom) are parties to the Convention on the Territorial
Sea and Contiguous Zone; six (Spain, Italy, Israel, Albania,
104

Yugoslavia and the United Kingdom) are parties to the High Seas Convention; and nine (Spain, France, Malta, Israel, Cyprus, Greece, Albania, Yugoslavia and the United Kingdom) are parties to the Continental Shelf Convention. Only three, (France, Spain and Yugoslavia) are parties to the Convention on Fishing and Conservation of the Living Resources of the High Seas.

It has already been pointed out, however, that non-membership of these Conventions is not necessarily significant in respect of provisions which are also rules of general customary international law.

In ratifying the Conventions, some of the above states have entered reservations. Thus Spain has entered a reservation that its accession to the three Conventions is not be interpreted as recognition of any rights or situations in connection with the waters of Gibraltar other than those referred to in Article 10 of the Treaty of Utrecht of 13 July 1713, between the Crowns of Spain and Great Britain. Spain has also entered a reservation to Article 1 of the Continental Shelf Convention that the existence of any accident of the surface, such as a depression or a channel, in a submerged zone shall not be deemed to constitute an interruption of the natural extension of the coastal territory into or under the sea. Italy entered a reservation to Article 24(1) of the Territorial Sea and Contiguous Zone Convention, which relates to the contiguous zone, reserving its right to exercise surveillance within the zone for the purpose of preventing and punishing infringements of the customs regulations in whatever point of this belt such infringements may be committed.

Unilateral practice. The majority of Mediterranean states claim a territorial sea of 12 nautical miles (see Table 2.6 page 51); it is likely that only Albania (15 miles) claims more than this. A minority of states claims less than 12 miles, including Greece and Israel (6 miles) and the United Kingdom in respect of its dependent territories (3 miles). Lebanon has not proclaimed any particular distance, while the Syrian claim is obscure in extent. Although France and Spain have each declared exclusive economic zones of 200 miles these have not yet been applied to the Mediterranean. Morocco has promulgated an economic zone of 200 miles for both the Atlantic and the Mediterranean.

105

Several states, including France, Greece, Spain, Malta and Italy, have issued continental shelf legislation. This legislation usually describes the limits of the shelf in terms of the "exploitability" criterion embodied in Article 1 of the 1958 Continental Shelf Convention, or, in the case of Malta, a median line in the absence of agreement with neighbouring states.

Several states including France, Spain, Egypt, Italy, Morocco, Yugoslavia and Turkey have instituted a system of straight baselines or single baselines for some part of their Mediterranean coast so as to enclose offshore islands or an indented coast. In 1973 Libya asserted a claim to the Gulf of Sirte north to latitude 32° 30' where the feature is about 300 miles wide. In 1981 this led to a confrontation with the United States.

Each of the Mediterranean states has enacted municipal legislation to control its maritime areas, including pollution from land and vessel-based sources. It is beyond the scope of this paper to discuss this legislation. In the words of one writer who has analysed it "perhaps the only generalisation that can be drawn from an examination of coastal state practice is that it has been uneven and fragmented in scope, purpose and application." (1)

Bilateral practice. In addition to the Treaty of Osimo of 1975 between Italy and Yugoslavia which came into force on 16 June 1978 and which described the maritime frontier in the Gulf of Trieste, only four other bilateral delimitation agreements, a small number in comparison to the number of potential maritime boundaries in the Mediterranean, have been concluded. Italy has been a party to all four, with Yugoslavia in 1968, Tunisia in 1971, Spain in 1974 and Greece in 1977. All relate to the continental shelf and may be classified as agreements between opposite, rather than adjacent states. A number of observations should be made about each of these delimitations.

1) Italy-Yugoslavia 1968 (Fig. 3.1 - in force 21 January 1970). The most anomalous feature of the area delimited consists of certain Yugoslav islands, Jabuka, Kajola and Pelagruz, situated in the Adriatic Sea about half way between the two land masses. If a strict equidistant line had been adopted the boundary would have therefore been drawn to the disadvantage of Italy. Under the agreement, the above islands were given a 12-mile territorial sea and the

106

area outside this zone, even though nearer to the Yugoslavia islands than to the Italian mainland, was allocated to Italy. As a balancing factor, the Italian island of Pianosa was not taken into account in establishing the boundary.

It is clear that the baseline on the Yugoslavian coast was established using the straight baseline system adopted by Yugoslavia in 1965 which encloses the chain of islands stretching along most of the length of its coast.

2) Italy-Tunisia 1971 (Fig. 3.2. - in force 6 December 1978). This follows a median line with some striking exceptions. These exceptions constitute the Italian islands of Panelleria, Lampedusa, Lampione, and Linosa which are located nearer to the Tunisian coast than to the Italian coast of Sicily. For the purposes of arriving at the median line the islands were disregarded but a zone of 13 miles (12 miles territorial sea and contiguous zone and 1 mile continental shelf) was allocated to each island. The islands were so situated that the median line constructed in disregard of them could be diverted to follow the 13-mile arcs without totally cutting off the islands from the main part of the Italian continental shelf.

3) Italy-Spain 1978 (Fig. 3.3 - in force 16 November 1978). This agreement provides for a boundary of some 137 nautical miles in length between Minorca and Sardinia. The boundary follows a median line but does not take account of the straight baseline systems adopted by Spain and Italy for their respective islands.

4) Italy-Greece 1977 (Fig. 3.4 - in force 12 November 1980). This agreement appears to follow a median line between the Greek and Italian coasts which takes into account the large islands from Corfu to Zante and, on the Italian side, a system of straight baselines closing the Gulf of Taranto and other shallower features on the east coast of Calabria.

Each of these agreements contains a provision that in the event of a deposit extending on both sides of the boundary the two parties should work together, after consulting the concession holders, with the aim of reaching agreement on the manner in which the deposit is to be exploited.

Current delimitation problems

1) Tunisia-Libya: By its judgement of 14 February 1982, the International Court of Justice indicated the principles and rules for the delimitation of the continental

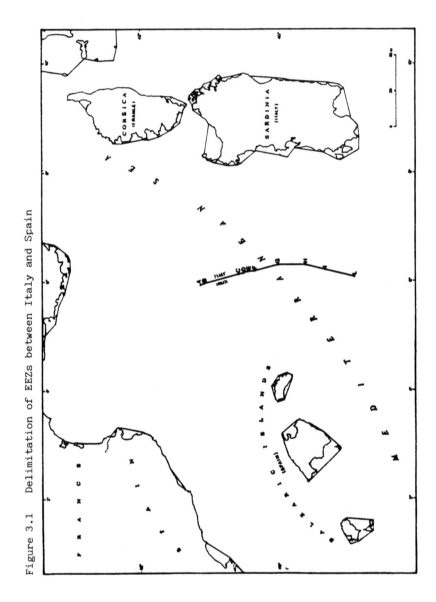

Figure 3.1   Delimitation of EEZs between Italy and Spain

Figure 3.2 Delimitation of EEZs between Italy and Tunisia

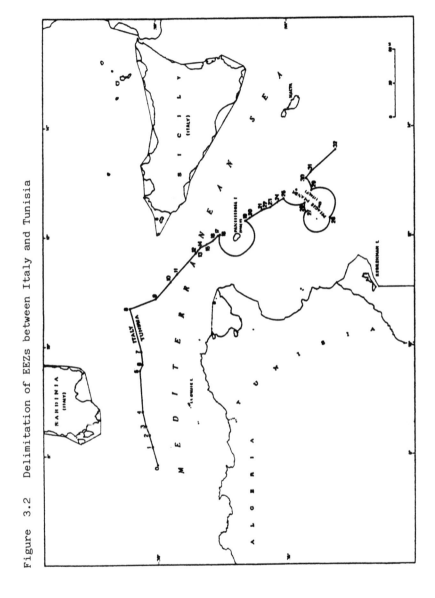

Figure 3.3   Delimitation of EEZs between Italy and Greece

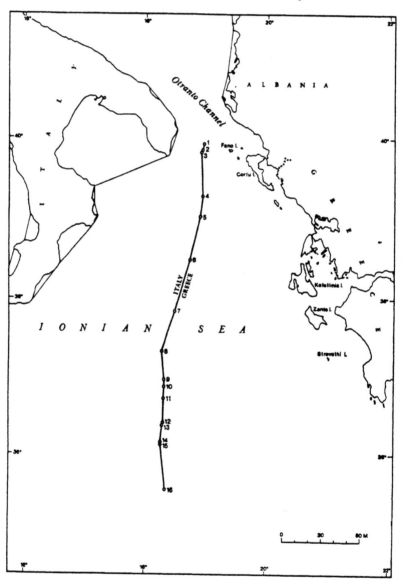

Figures 3.1 to 3.4
Source: B. Conforti and G. Francalanci, eds.,
ATLAS OF THE SEABED BOUNDARIES, Giuffre, Milan, 1979.

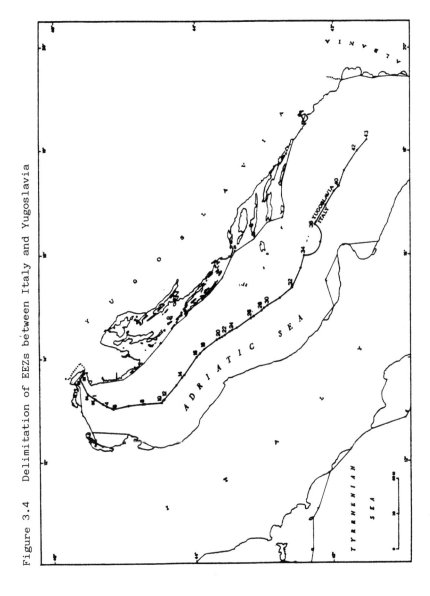

Figure 3.4   Delimitation of EEZs between Italy and Yugoslavia

shelf between Tunisia and Libya and clarified the practical method for the application of these rules and principles, so as to enable the experts of the two countries to draw a line of delimitation. In arriving at its conclusions, the Court considered itself obliged to effect the delimitation in accordance with equitable principles and taking account of all relevant circumstances. The Court held that as the area in question formed a single continental shelf, and was the natural prolongation of the landmass of both parties, no criterion for delimitation could be derived from the principle of natural prolongation. The relevant circumstances found by the Court were the general configuration of that part of the coasts which in the Court's view was the area relevant to the delimitation, particularly the change in direction of the Tunisian coastline in the Gulf of Gabes; the existence and position of the Tunisian Kerkennah islands; the land frontier between the parties and their respective practice with regard to the granting of petroleum concessions offshore the land frontier; and an element of a reasonable degree of proportionality between the extent of the continental shelf areas and the length of the relevant part of the coast to which they appertain.

2) <u>Libya-Malta</u>: The International Court of Justice has been requested by these two states to delimit the continental shelf boundary between them.

3) <u>Greece-Turkey</u>: There are about 1,000 islands, mainly Greek, in the Aegean Sea, including some Greek islands which are close to the mainland coast of Turkey. If, as Greece claims, each island generates a continental shelf as well as a territorial sea, a large part of the seabed on the Turkish side of a hypothetical median line drawn between the two mainlands is claimed by Greece. In 1974 Turkey issued a decree which proclaimed the western edge of the Turkish continental shelf to be a median line drawn between the two mainlands, ignoring the islands. This has the effect of enclaving the Greek islands situated to the east of the median line, to which Turkey concedes only the 6 miles of territorial sea claimed by both states, and no continental shelf. Turkey maintains that the Greek islands in question are located on Turkey's continental shelf this being the "natural prolongation" of the Turkish land mass. An attempt by Greece to refer the dispute to the International Court of Justice was unsuccessful.

COOPERATION BETWEEN MEDITERRANEAN STATES

The General Fisheries Council for the Mediterranean.
This is an organisation formed in 1949 under the auspices of
the FAO amongst governments "having a mutual interest in the
development and proper utilisation of the resources of the
Mediterranean and contiguous zones." The riparian Mediterranean states, with the exception of Albania and Israel, are
members, as are Romania and Bulgaria. The functions of the
organisation are set out in Article 4 of the Agreement as
follows:

    a. To formulate all oceanographical and technical
       aspects of the problems of development and
       proper utilisation of aquatic resources;
    b. To encourage and co-ordinate research and the
       application of improved methods employed in
       fishery and allied industries with a view to
       the utilisation of aquatic resources;
    c. To assemble, publish, or disseminate all
       oceanographical and technical information
       relating to aquatic resources;
    d. To recommend to Members such national and
       international research and development
       projects as may appear necessary or desirable
       to fill gaps in such knowledge;
    e. To undertake, where appropriate, cooperative
       research and development projects directed to
       this end;
    f. To propose, and where necessary to adopt,
       measures to bring about the standardisation of
       scientific equipment, techniques, and nomen-
       clature;
    g. To make comparative studies of fishery legis-
       lation with a view to making recommendations
       to its Members respecting the greatest
       possible co-ordination;
    h. To encourage research into the hygiene and
       prevention of occupational diseases of fisher-
       men;
    i. To extend its good offices in assisting
       Member(s) to secure essential materials and
       equipments;
    j. To report upon such questions relating to all

oceanographical and technical problems as may
be recommended to it by Members or by the
Organisation and if it thinks proper to do so,
by other international, national, or private
organisations with related interests;

k. To transmit every two years, to the Director-
General of the Organisation, a report embody-
ing its views, recommendations and decisions,
and make such other reports to the Director-
General of the Organisation as may seem necessary or desirable.

The International Commission for the Conservation of
Atlantic Tunas. This was established in 1966 under a Convention which defines the area in question as including all
seas adjacent to the Atlantic Ocean. Only three Mediterranean countries, France, Spain, and Morocco are members.
The aim of the Commission is to maintain the tuna population
at levels which will permit the maximum sustainable catch.
To this end there is coordination, promotion and publication
of research.

The Work of the United Nations Environmental
Programme. In early 1975, the United Nations Environmental
Programme organised in Barcelona an inter-governmental meeting on the protection of the Mediterranean. The meeting
adopted an Action Plan for the protection and development of
the Mediterranean Basin. In particular, the Plan envisaged
the integrated planning of the development and management of
the resources of the Basin, the coordination of pollution
monitoring and research in the Mediterranean, and the conclusion of a framework convention with related protocols and
technical annexes. At the same time the coastal states were
exhorted to become parties to the 1973 International Convention on the Prevention of Pollution from Ships (The
London Convention) and to use their efforts within the
International Maritime Consultative Organisation to have the
Mediterranean designated as a special area for the purposes
of Annex II of that Convention.

The adoption of the Action Plan led to the convening
of a Conference in Barcelona in February 1976 which was
attended by 16 Mediterranean States. The Conference adopted
three instruments: a Convention for the Protection of the
Mediterranean Sea against Polution, a Protocol for the Prevention of Pollution of the Mediterranean Sea by Dumping

from Ships and Aircraft, and a Protocol concerning Cooperation in Combating Pollution of the Mediterranean Sea by Oil and other Harmful Substances in Cases of Emergency.

The Convention defines the "Mediterranean Sea Area" as "the maritime waters of the Mediterranean Sea proper, including its gulfs and seas, bounded to the west by the Meridian passing through Cape Spartel lighthouse, at the entrance of the Straits of Gibraltar, and to the east by the southern limits of the Straits of the Dardanelles between Mehmetcik and Kumlake lighthouses".

The internal waters of the contracting parties are expressly excluded from the Area.

In the preambles to the Convention, the contracting parties, "conscious of the economic, social, health and cultural value of the marine environment of the Mediterranean Sea Area" declare their full awareness "of their responsibility to preserve this common heritage for the benefit and enjoyment of present and future generations". This "common heritage" is clearly not analogous to the "common heritage of mankind" provided for in respect of the deep seabed since the Convention goes on to preserve from prejudice "the present or future claims and legal views of any State concerning the law of the sea and the nature and extent of coastal and flag State jurisdiction".

Article 4 sets out the "General Undertakings".

The contracting parties undertake thereunder to take measures to prevent, abate and combat four kinds of pollution (i) pollution caused by dumping from ships and aircraft, (ii) pollution from ships' discharges, other than dumping, (iii) pollution resulting from the exploration and exploitation "of the continental shelf and the seabed and subsoil", (iv) pollution from rivers, coastal establishments or outfalls or other land-based sources within their territories.

Cooperation is sought (i) in taking the necessary measures in dealing with pollution emergencies in the Area, (ii) in monitoring pollution, (iii) in scientific and technological cooperation. At the same time the contracting parties undertake to cooperate as soon as possible to formulate and adopt appropriate procedures for the determination of liability and compensation or damage resulting from the pollution of the marine environment deriving from violations of the provisions of the Convention and Protocols.

The two Protocols take matters further. The Dumping Protocol is based on the Oslo and London Conventions of 1972, though it is stricter than the regimes laid down in those instruments. The Emergency Protocol provides that in cases of grave and imminent danger to the marine environment, the coast or related interests of one or more of the parties due to the presence of oil or other harmful substances, the contracting parties shall cooperate in taking the necessary measures.

The Convention and associated Protocols entered into force on 12 February 1978.

Following a Conference held in Athens in May 1980, under the auspices of UNEP, a Protocol for the Protection of the Mediterranean Sea from Land-Based Sources was adopted. This Protocol obliges the contracting parties to take all appropriate measures to prevent, abate, combat and control pollution of the Mediterranean Sea Area caused by discharges from rivers, coastal establishments or outfalls, or emanating from any other land-based sources within their territories. The Area to which the Protocol applies is not only the Mediterranean Sea Area defined in the 1976 Convention but includes waters on the landward side of the baselines from which the breadth of the territorial sea is measured extending, in the case of watercourses, up to the freshwater limit. The area also includes saltwater marshes communicating with the sea.

Sub-regional agreements. Italy has concluded two agreements with its neighbours which might serve as precedents for other cooperative ventures in the Mediterranean.

1) Italy-Yugoslavia: An agreement for collaboration in safeguarding the waters of the Adriatic and its coastal zones from pollution was signed in 1974 and came into force in 1977. Its main aim is to create an institutional mechanism for collaboration between the two states, in order to assess and control pollution of the area, including pollution from land-based sources.

The agreement establishes a Joint Commission, nominated by the two states, assisted by a sub-commission of scientists and experts. Its aims are: (a) to assess the problems concerning pollution in the area; (b) to propose and recommend research programmes to the two states; (c) to evaluate bilateral programmes and provide for their coordination; (d) to propose to the two states necessary measures for combating existing pollution and preventing future

116

pollution; (e) to suggest to the two states programmes of international regulations in order to ensure the purity of the waters of the Adriatic. The decisions of the Commission are taken by unanimity.

2) Italy-Greece; An agreement on the protection of the marine environment of the Ionian Sea and its coastal zones was signed in 1979. The aims are similar to those of the Adriatic agreement. It was adopted on the basis of "the spirit of co-operation upheld by the Parties to the Barcelona Convention". Under the agreement Italy and Greece are bound "to cooperate to prevent, combat and gradually eliminate" the pollution of the Ionian Sea and its coastal zones. The decisions of the Commission are taken by unanimity.

THE LEGAL ASPECTS OF FUTURE DELIMITATIONS IN THE MEDITER-RANEAN

Under customary international law as well as under the 1959 Convention on the Continental Shelf every coastal state is entitled ipso jure to the shelf appurtenant to its land mass. This entitlement does not depend on proclamation or occupation. Each Mediterranean state thus has a continental shelf irrespective of proclamation or whether its lateral or opposite boundaries have been delimited. In the North Sea Continental Shelf Cases, the International Court of Justice stated (paragraph 39): "From this notion of appurtenance is derived the view which, as has already been indicated, the Court accepts, that the coastal State's rights exist ipso facto and ab initio without there being any question of having to make good a claim to the area concerned, or of any apportionment of the continental shelf between different States."

The definition of the continental shelf under the 1982 Convention, not necessarily being linked to physical realities, would cover the entire bed and subsoil of the Mediterranean, leaving no part to the "common heritage of mankind" regime provided for the seabed beyond national jurisdiction. Furthermore, the 200-mile exclusive economic zone concept would cover the entire water area of the Mediterranean, outside internal and territorial waters. The conclusion must therefore be drawn that actually or potentially the whole Mediterranean Sea is subject to some form of coastal state jurisdiction.

There is nothing in the 1982 Convention any more than in existing customary and treaty law which prevents or even impedes a Mediterranean state from claiming a share of the Mediterranean in the form of internal waters, territorial sea, continental shelf and exclusive economic zone. Article 123 of the 1982 Convention requires the coastal states of a semi-enclosed sea to endeavour to coordinate the management and exploitation of the living resources of the sea, but nothing is said of the non-living resources such as minerals. The practice of Mediterranean states in negotiating continental shelf boundary agreements or of litigating such boundaries with their neighbours indicates that it is unlikely that there will be amoratorium on continental shelf delimitations in the Mediterranean or that a scheme for joint management of the shelf is near.

With regard to the natural resources of the waters, as opposed to the subjacent lands, there is some reason to think that the position might evolve otherwise. Article 123, as stated above, does refer to the living resources of the sea. Furthermore, as there is nothing naturally appurtenant about a 200-mile zone of water, states might not consider that they already have such a zone by the principle of _ipso jure_ appurtenance. If Spain enters the European Community, a large part of the waters of the western Mediterranean would become potentially Community waters in which national rights of fishing will be diminished in favour of Community use. There is thus some cause to believe that Mediterranean states might refrain from claiming national exclusive economic or fisheries zones of substantial breadth. At present EC states exclude all foreign fishing from a six-mile zone.

If the Mediterranean states do enter into negotiations or litigation with a view to the delimitation of the Mediterranean - and we have seen that there is every likelihood they will continue to do so for the continental shelf - on which principles will future delimitations be based?

It has already been seen that in respect of opposite or adjacent territorial sea, including the subjacent lands and the superjacent airspace, the basic principle, failing agreement, is one of equidistance. This is set out in Article 12(1) of the 1958 Convention and in Article 15 of the 1982 Convention.

In respect of the continental shelf, however, the International Court of Justice in the <u>North Sea Continental Shelf Cases</u> in 1969 concluded that "the notion of equidistance as being logically necessary, in the sense of being an inescapable <u>a priori</u> accompaniment of basic continental shelf doctrine, is incorrect". In the more recent <u>Tunisia/Libya Case</u> the same Court stated that "equidistance is not, in the view of the Court, either a mandatory legal principle or a method having some privileged status in relation to other methods". In that case, neither Tunisia nor Libya had in fact relied on equidistance.

In the <u>North Sea Continental Shelf Cases</u> in 1969 the Court stated the basic principle of shelf delimitation as follows (paragraph 90): "... the international law of continental shelf delimitation does not involve any imperative rule and permits resort to various principles or methods, as may be appropriate, or a combination of them, provided that, by the application of equitable principles, a reasonable result is arrived at."

In the <u>Tunisia/Libya Case</u> in 1982 the Court stated (paragraph 71): "(The Court) ... is bound to apply equitable principles as part of international law, and to balance up the various considerations which it regards as relevant in order to produce an equitable result."

In the <u>North Sea Cases</u> the Court listed as factors to be taken into account the general configuration of the coasts of the parties including the presence of special or unusual features, the physical and geographical structure, and a reasonable degree of proportionality between the lengths of the respective coasts and the area of shelf appertaining thereto.

In the <u>Tunisia/Libya Case</u>, the Court rejected an argument based on the overriding importance of "natural prolongation" and ruled that the shelf in dispute was the natural prolongation of the land mass of both parties, i.e. there was an overlap of natural prolongations. It was unwilling to consider that geological factors were paramount in assessing natural prolongation. The Court thus gave a greater weight to the geographical features of the coastline than to the geology of the submerged lands. This approach may be relevant in the Aegean where Turkey claims <u>inter alia</u> that the shelf is a "natural prolongation" of its land mass, but it cannot be concluded therefrom that geological factors will be subservient to geographical factors in every future delimitation. By permitting a state a juridical continental

119

shelf of 200 nautical miles from the baseline of the territorial sea when that state does not have a physical continental margin, the new definition of continental shelf in the 1982 Convention seems also to be turning away from natural prolongation. There is nothing natural about a fictitious shelf extending to a uniform distance. Algeria and Morocco, for example, might be beneficiaries of this concept. In delimitations between adjacent states of this kind equidistance might be more important than it was in the North Sea and Tunisia/Libya Cases. Similarly in future delimitations of exclusive economic zones as contrasted with the continental shelf equidistance may still be relevant. It may also be important in delimitations between opposite as contrasted with adjacent states.

In the Tunisia/Libya Case, both parties relied on economic factors as relevant to the delimitation process. Thus Tunisia argued that its relative poverty in comparison to Libya should be taken account of. Libya, while denying that economic poverty should be relevant, argued that the productivity of oil and gas wells on the respective areas of shelf was a relevant factor. The Court rejected the Tunisian argument but gave some weight to the Libyan submission. The relevant paragraph (paragraph 107) from the Court's judgement reads as follows:

> ... these economic considerations cannot be taken into account for the delimitation of the continental shelf areas appertaining to each Party. They are virtually extraneous factors since they are variables which unpredictable national fortune and calamity, as the case may be, might at any time cause to tilt the scale one way or the other. A country might be poor today and become rich tomorrow as a result of an event such as the discovery of a valuable economic resource. As to the presence of oil wells in an area to be delimited, it may, depending on the facts, be an element to be taken into account in the proceeds of weighing all relevant factors to achieve an equitable result.

The unpredictable variables mentioned above were discussed by Judge Evensen, the Norwegian ad hoc judge for Tunisia in the case. In his dissenting judgment Judge Evensen put forward the proposal of joint exploitation for a restricted area of overlapping claims. With so many potential maritime boundary disputes in the Mediterranean this could provide a possible future solution for some of them.

Judge Evensen was probably influenced by his membership in 1981 of a Conciliation Commission set up by Iceland and Norway to make recommendations for the dividing line of the continental shelf between Iceland and the Norwegian island of Jan Mayen, some 290 miles distant from Iceland.

Iceland had proclaimed a 200-mile exclusive economic zone and claimed that it was entitled to a shelf even beyond this limit as a natural prolongation of the land mass of Iceland. The Commission was instructed to "take into account Iceland's strong economic interests in these sea areas, the existing geographical and geological factors and other special circumstances". Having come to the conclusion that the submarine area between Iceland and Jan Mayen was a "micro-continent" and not a natural prolongation of either, the Commission defined an area, nearly three-quarters of which lay on the Jan Mayen side of the Icelandic 200-mile limit. In this area the Commission proposed joint development. Within that part of the area which fell inside the 200-mile line, Icelandic legislation, oil policy and control would apply, with Norway having the right to acquire up to a 25% stake in any joint venture. Within that part of the area which fell outside the Icelandic limit, Norwegian legislation etc. would apply with Iceland having the right to acquire up to a 25% stake in any joint venture.

Finally, the effect on military strategy of the possible partition of the Mediterranean amongst the coastal states should be briefly discussed. As shown above, the 1982 Convention contains a new concept of "transit passage" applicable to straits of less than 24 miles wide. Although it is probable that the Strait of Gibraltar will fall under this regime, transit passage is certainly not yet in force as a rule of treaty law or customary law. Indeed, the Spanish delegate at the beginning of UNCLOS III made a speech in opposition to the proposal to replace innocent passage as the regime in force in straits. He stated: "Straits used for international navigation were an integral part of the territorial sea in so far as they lay within territorial waters. Any attempt to set up separate regimes for the territorial sea and for straits would clearly violate the fundamental principle of sovereignty of the coastal State over its territorial sea... "

Professor John Norton Moore (2) has summarised the ambiguities and inadequacies of "innocent passage" as applied to straits by the 1958 Convention on the Territorial Sea and Contiguous Zone as follows:

(a) failure to recognise the different community policies of a regime of passage through the territorial sea and a regime of transit through straits used for international navigation;

(b) no right of overflight over the territorial sea;

(c) submarines in innocent passage must navigate on the surface and show their flag;

(d) subjectivity inherent in the definition of "innocent passage" coupled with the right of the coastal State to take the necessary steps in its territorial sea to prevent passage which is not innocent;

(e) uncertain and unbalanced coastal State regulatory competence over vessels in innocent passage, particularly the uncertain prescriptive and enforcement competence for dealing with vessel-source pollution;

(f) uncertainty over which straits are those "used for international navigation";

(g) failure of some strait States to adhere to the 1958 Convention and the consequent assertion of even more restrictive rules such as the requirement for prior notification for the transit of warships, restrictive passage through "archipelagic waters" or broadly defined "historic waters".

Some of these ambiguities remain under the 1982 Convention's proposals. In particular, it is unlikely that straits in the Mediterranean other than Gibraltar will be subject to transit passage, since it can be argued that there are alternative routes of similar convenience.

An important feature of the 1982 Convention for the future of military strategy in the Mediterranean is the inclusion of a maximum breadth of 12 nautical miles for the territorial sea. It is probable that a customary rule of international law has developed permitting states to extend their territorial sea to this distance so that irrespective of the fate of the 1982 Convention such a distance is already lawful. This development will result, and already has resulted, in areas of sea being unavailable for military use except with the consent of the coastal state. It is significant, however, that 12 miles is stated to be the maximum breadth of the territorial sea, thus preventing even more substantial claims.

The concept of the exclusive economic zone, though preserving the high seas freedoms of navigation and over-flight by Article 58(1) of the 1982 Convention, may cause some erosion of the exercise of these rights because of the coastal state's exclusive right to construct artificial islands, installations and structures in the zone. Article 59 provides that any conflict between the interests of the coastal state and other states "should be resolved on the basis of equity and in the light of all the relevant circumstances, taking into account the respective importance of the interests involved to the parties as well as to the international community as a whole".

A multilateral convention relevant to the problem is the 1970 Treaty on the Prohibition of the Emplacement of Nuclear Weapons and other Weapons of Mass Destruction on the Seabed and the Ocean Floor and in the Subsoil thereof, passed by 104 votes to 2, with 2 abstentions, and in force on 18 May 1972. The Treaty has over 60 Parties.

Article I of the Treaty provides that:

(1) The States Parties to this Treaty undertake not to emplant or emplace on the sea-bed or ocean floor and in the subsoil thereof beyond the outer limit of a sea-bed zone as defined in Article II any nuclear weapons or any other types of weapons of mass destruction as well as structures, launching installations or any other facilities specifically designed for storing, testing or using such weapons.

(2) The undertakings of paragraph 1 of this Article shall also apply to the sea-bed zone referred to in the same paragraph, except

that within such sea-bed zone, they shall not apply either to the coastal State or to the seabed beneath its territorial waters.

(3) The States Parties to this Treaty undertake not to assist, encourage or induce any State to carry out activities referred to in paragraph 1 of this Article and not to participate in any other way in such actions.

Article II of the Treaty reads: "For the purpose of this Treaty the outer limit of the sea-bed zone referred to in Article 1 shall be coterminous with the 12-mile outer limit of the zone referred to in Part II of the Convention on the Territorial Sea and the Contiguous Zone (i.e. the contiguous zone)..."

Although there were some early proposals in UNCLOS III to prohibit the construction or operation of military installations or devices on or over the continental shelf without the consent of the coastal state, the 1982 Convention does not contain any such specific provision.

NOTES

1. Scott G. Truver, <u>The Strait of Gibraltar and the Mediterranean</u>, 1980, p. 123.

2. John Norton Moore, <u>American Journal of International Law</u>, Vol. 74, pp. 85-86.

Chapter Four

THE ARAB STATES AND MAJOR SEA ISSUES

Nazih N.M. Ayubi

Arab countries, like those of the rest of the Third World, are no longer mere outsiders or simple spectators in setting the agenda for sea policies. They have not been as influential as, for instance, the Latin American countries were in the recent process of reshaping sea policies that resulted in the adoption by the United Nations of the new and comprehensive Law of the Sea. However, the very fact that practically all Arab states - many of which are quite new even in their acquisition of statehood and most of which have not previously participated in similar conferences - played an active role in the eight-year-long activities of the Third United Nations Conference on the Law of the Sea (UNCLOS III), is by itself a considerable achievement.

ARAB STATES AND THE THIRD UNCLOS

The active participation of Arab and Third World countries has emphasised that "oceans' politics" no longer revolve around the simple issues of navigation and fisheries in a "free seas regime". The oceans agenda has become far more reflective of the increasingly complex interdependence among nations, and has become strongly influenced by various economic and technological changes that have raised a number of distributional questions. The major new issue is deep seabed resources, whose potential has increased the number of countries interested in the sea issue far beyond those who are major users of the oceans to include the coastal and less developed countries. (1) The high-technology issues of deep sea mining and research have also tended to highlight a cleavage over sea policies between the developed and the developing countries (known as the Group of 77), a cleavage that has often been reinforced through the channels of the existing regional groupings such as the Organization for African Unity or the League of Arab States. (2)

For the Middle Eastern countries, in particular, the significance of the Law of the Sea is understandable, given their strategic location at the meeting points of some of the busiest waterways of the world: the Mediterranean Sea, the Red Sea and the Persian Gulf. It is also interesting to note that of the four countries that opposed the new treaty, two countries are Middle Eastern (Turkey and Israel), while a third (the United States of America) was recently involved in a sea quarrel with a Middle Eastern country (Libya).

Turkey has an outstanding conflict with Greece over territorial waters in the Aegean Sea, and with the present legalisation of the 12-mile territorial waters rule, Turkey is trying to attract the political support of other Middle Eastern countries in its attempt to resist "by all necessary measures" the extension of Greek territorial waters to the new internationally acknowledged limit. (3)

Israel has a sensitive position in the Red Sea (to be considered below), which may at least partly explain its rejection of the new law.

As for the United States, the crux of the American objection lies in the Reagan Administration's view that the treaty does not adequately protect the American firms that have pioneered the technology and the exploration for the mineral nodules (particularly manganese but also cobalt, zinc, copper and others) that can be scooped from the deep seabed. Among the interesting points denounced by the United States in the draft convention is that it "contains provisions concerning liberation movements, like the PLO, and their eligibility to obtain a share of the revenues of the Seabed Authority. (4)

The chairman of the Group of 77 expressed the feeling of the majority of states, Arab countries included, when he said that "the United States government cannot reject the work of over 150 nations including its own predecessor government for almost a decade, for in doing so it would be destroying the principle of good faith negotiations. There have been scores of changes in regimes in different countries since the work on the treaty was started, but no new regime has so far disowned what its predecessors had striven to achieve in the field of international cooperation... " (5)

127

The Arab States and Major Sea Issues

The United States government, however, voted against the agreement, and on 9 July 1982 President Reagan announced that America was not going to sign, thus leaving in limbo ultimate control over trillions of dollars worth of minerals waiting to be mined from the seabed. The treaty was nevertheless opened for signature in December 1982 so that it would come into force when sixty countries had signed it. (6) Even so, it would seem that a significant opportunity for international sea management may be lost, and all the more so since the Law of the Sea and the submitted Convention deal with much more than "the philosophy of regulating manganese nodule mining: they deal with military and commercial navigation, overflight and communications, fisheries, continental shelf gas and oil, prevention of pollution, marine scientific research, and settlement of disputes... " (7)

Before proceeding any further, therefore, let us consider some of the general attitudes of Middle Eastern countries towards some of these issues.

GENERAL ATTITUDES AND POSITIONS

Although Iran, Saudi Arabia, Oman, South Yemen and Qatar have made recent claims to exclusive rights (essentially for control of fishing activities) over areas of the sea much larger than those previously contemplated by international rules, it may be noted that, unlike other states, most Middle Eastern countries have abstained from making jurisdictional claims of considerable dimensions. (8) None of them claims a territorial sea in excess of twelve miles, and only very few of them claim exclusive fishing zones in excess of their territorial waters. Domestic legislations of these states also generally conform with the internationally recognised guidelines regarding the continental shelf.

In fact, Middle Eastern countries have had comparatively less influence on the recent process of reshaping the international law of the sea than have some other developing countries. Their position with regard to many of the issues discussed at the Third Conference were, on the whole, developed in response or as a reaction to trends by other states.

Like most other developing countries, Arab countries have no great strategic interests apart from their economic concerns. Their interests place them generally within the

128

"coastal" group of states, rather than within the "maritime" group that is not only already active in navigation but also has the highest potential for exploiting seabed resources.

The attitudes of almost all Arab states towards the main elements contained in the Law of the Sea is fairly standard. There is a general acceptance of the twelve-mile territorial sea limit, Somalia and Mauritania being the only members of the League of Arab States to claim territorial seas in excess of twelve miles. (9) As far as straits are concerned, some Middle Eastern countries, especially those bordering on straits, support the principle of non-suspendable innocent passage for foreign ships, particularly warships; whereas others, especially Iraq, support the newly introduced principle of "free" or "transit" passage. The majority of them, however, insist that the "regime of straits" should be strictly confined to straits which connect two parts of the high seas. (10)

As for the 200-mile exclusive economic zone, the concept is acceptable to practically all Middle Eastern states, with the exception of Kuwait which has developed a distant-water fishing fleet that fishes off the coasts of Africa and in the Indian Ocean. Owing, however, to the variance of interests among the different states, there was no agreement among Middle Eatern countries in the Third UNCLOS as to whether the Mediterranean, the Red Sea and the Persian Gulf should be regarded as "enclosed or semi-enclosed seas", especially in connection with the exclusive economic zone concept, but also in relation to the freedom of navigation.

Egypt - whose position is quite typical - has taken the view that within the 200-mile exclusive zone, the coastal states must observe and enforce international standards, especially with regard to shipping and seabed mining. It has recognised, however, the need for special arrangements between coastal states whose area of national jurisdiction could not extend for 200 miles without overlapping, a situation which occurs in the Red Sea and the Mediterranean (on both of which Egypt has borders).

To many Middle Eastern countries, the questions relating to bays and navigation through straits are of special significance, since their coasts tend to be marked by many bays and embrace important international straits. In the case of Egypt, the controversy surrounding the legal status of, and the rules governing the right of passage through, the Gulf of Aqaba and the Straits of Tiran, have been par-

ticularly significant. The Egyptian position is that the
Gulf and the Straits do not constitute international waterways or connote high seas, but that they are simply territorial waters in which there is only a right of passage for
innocent ships. Saudi Arabia, also with an eye on Israel's
illegal presence in the Gulf of Aqaba, maintains that in
straits "'passage is innocent unless it is prejudicial to
the security of the coastal State. Such passage is not
innocent when it is contrary to the present rules or to
other rules of international law". (11)

As far as exploration and exploitation of the seabed
and ocean floor and the subsoil thereof beyond the limits of
national jurisdiction are concerned, the Arab states, like
most developing countries, believe that the proposed international authority for organising these activities should
have strong functional and supervisory powers, and should be
essentially an operating rather than a licensing authority.
(12)

Relations among Middle Eastern countries themselves
are likely to be influenced by the various provisions of the
new treaty. Although several Arab countries had already
announced - in advance of the treaty - a 12-nautical-mile
limit to their territorial waters, and while several had
grabbed the 200-mile "exclusive economic zones" of sea that
the new code is to legalise, the adoption of the new sea
package is not by itself likely to end immediately all the
outstanding "sea conflicts" between them, of which there are
indeed quite a few. It would be useful therefore to consider
briefly at this juncture some of the most important of the
outstanding issues involving Middle Eastern countries in the
Mediterranean, the Red Sea and the Persian Gulf.

COOPERATION AND CONFLICT IN THE MEDITERRANEAN SEA

Arab states occupy the eastern and the southern shores
of this median sea which they call the "White Sea" (al-bahr
al-abyad). Owing to the geographical conformation of the
Mediterranean and to the various political interests and
legal claims by its states, there are some controversies
over the delineation of territorial waters, especially with
regard to bays and gulfs. Egypt has indicated five bays as
falling within its territorial seas even though they are
"bays of considrable breadth and relatively small depth".
(13)

Libya also considers the Gulf of Sirte (Sidra) as part of its territorial waters, and this claim led to the development of a bloody confrontation with the United States in 1981 - ironically, in fact, since it occurred as the resumed tenth session of the Third UNCLOS was convening. In August 1981, two Libyan planes were shot down by United States navy planes in a dogfight that took place over the Gulf of Sidra, an area that Libya regards as its territorial waters while the United States considers it international waters. President Reagan announced that the exercise was staged "because we could not go on recognising a violation of international waters". (14)

Many observers could see, however, that although based on a legal disagreement, this was obviously a political conflict since shooting down other countries' aircraft is not exactly the way to prove the freedom of the sea. Not surprisingly, therefore, even the Arabs who had no sympathy for Qadhafi at all condemned this action most strongly. The leaders of the Gulf Cooperation Council, for example, described the American action as "a provocative trap and mediaeval piracy in the high seas" that would only encourage "cowboy politics", (15) while one Egyptian made the comment that "Reagan now looks quite as mad as Qadhafi". (16) For months   afterwards,   American   authorities   expected   a retaliatory attack on Reagan, and since then both diplomatic and economic relations between the two countries have deteriorated seriously.

In fact, although the Gulf of Sidra does not meet the semi-circularity test or the twenty-four nautical miles closing limit required of a legal bay, these requirements do not apply to so-called "historic" bays, and it appears that Libya has based its claim over the Gulf both on historic considerations and on the principle of vital bays. In a declaration made on 9 October 1973; the Libyan Arab Republic announced that the Gulf of Surt or Sirte:

> ... constitutes an integral part of the territory
> of the Libyan Arab Republic and is under its
> complete sovereignty ... it constitutes internal
> waters, beyond which the territorial waters of
> the Libyan Republic start ... Through history and
> without any dispute, the Libyan Arab Republic has
> exercised its sovereignty over the Gulf. Because
> of the Gulf's geographic location commanding a

131

view of the southern part of the country, it is therefore crucial to the security of the Libyan Arab Republic. Consequently, complete surveillance over its area is necessary to ensure the security and safety of the State.

The declaration stated, furthermore, that private and public foreign ships were not allowed to enter the Gulf without prior permission from the Libyan authorities and in accordance with the regulations established by them in this regard. (17)

In a protest dated 11 February 1974 against Libya's declaration, the United States described the Libyan claim as "unacceptable as a violation of international law". It noted that the body of water in question could not be regarded as the juridical internal or territorial waters of Libya, nor did the Gulf of Sirte meet the standards of past open, notorious and effective exercise of authority, continuous exercise of authority, and acquiescence of foreign nations necessary to be regarded historically as Libyan internal and territorial waters. (18)

The revival of this legalistic conflict in the summer of 1981 was obviously motivated by political reasons. Indeed, the Libyans could see that some action was in the offing and in a note to the United Nations Security Council the Libyan Bureau for Foreign Liaison protested on 4 August 1981 that "the U.S. government has been escalating its campaign" against Libya in preparation for "a hostile action". (19) Two weeks later, on 18 August, a statement condemning an American naval exercise within the area claimed by Libya as territorial waters was issued by the Libyan government, in which the exercise was termed "uncalled for interference and provocation". The following day, on 19 August, United States F-14 fighter planes shot down two Libyan planes as they were carrying out military exercises 60 miles off the Libyan coast over the Gulf of Sidra. The fighters were patrolling a sixteen-ship navy task force. (20)

Following the air clash, Libya delivered a note of protest accusing the United States of "international terrorism". Both President Reagan and Secretary of State Alexander Haig denied that American naval exercises in the Gulf had been provocative. (21) One would hope that this really was the case, and that the American government, which has rejected the recent law of the sea, has not actually decided that force is more useful at sea than law, the former being always a strong temptation for a superpower. (22)

Most recently, the eastern Mediterranean has been the theatre for devastating turmoil, with the launching by Israel of its invasion of Lebanon in June 1982. An important part of the Israeli military operations, in addition to its blockade of food and medical supplies to West Beirut, were conducted off the Lebanese coast. After the fighting cooled down, there were also complaints from some Lebanese that the Israelis were exploiting their dominant position at sea by restricting the activities of Lebanese fishermen and by dumping on the Lebanese markets – often through "illegal" ports of entry – Israeli products on which customs duties were not being paid to the Lebanese authorities. (23)

Among other less explosive, but still conflictive, issues in the Mediterranean is the fact that the continental shelf boundaries in the central part of the Sea between Libya, Tunisia, Italy and Malta have not yet been determined. This problem has become more urgent because oil companies exploring the central Mediterranean have already made successful discoveries in Tunisian and Libyan areas, and it is feared that some oil may be found in areas claimed by two or more of these states. Discussions between Libya and Malta over oil exploration rights have not been successful, and the two countries have agreed to refer their dispute to the International Court of Justice. (24)

Exploitation of seabed resources does not represent a problem yet since exploration efforts seem so far to indicate that the Mediterranean is void of the valuable manganese nodules. However, fishing activities are likely to represent an increasingly difficult issue, as overfishing continues in this relatively small sea. Problems between, say, Greece and Egypt in the eastern Mediterranean and between the Iberian countries and Morocco and Mauritania in the western Mediterranean are likely to be of some significance.

In other areas there has been more cooperation between the states of the northern and those of the southern Mediterranean. For example, the Convention and Protocols for the Protection of the Mediterranean Against Pollution, adopted in 1976, were signed by Cyprus, France, Greece, Italy, Malta, Spain and Turkey, and from among Arab countries by Egypt, Lebanon and Morocco. (25)

The Transmed gas pipeline is also an important symbol of the growing functional cooperation between the European

133

shore and the Arab shore of the Mediterranean. This gas
pipeline has already been completed, linking Algeria to
Sicily via Tunisia, and the price agreement between Italy
and Algeria was signed in 1982. The agreement allows Algerian gas deliveries to Italy of about 12 billion cubic metres annually for a period of twenty-five years. (26)

THE RED SEA: AN ARAB LAKE?

Since the Red Sea is almost an Arab lake, it is understandable that there will be problems surrounding Israel's
position there. Over the years new and serious contending
concerns and strategies have emerged in such a manner as to
make the Red Sea a source of further Arab-Israeli conflict.
(27)

To make the Red Sea into an "Arab lake" was an idea
first broached by some Egyptian writers in the late fifties,
but it was not welcomed by the Saudis, who regarded
Nasserist policies as hostile to their interests. However,
after Nasser's death, the Saudis moved towards adopting this
policy, especially in view of the escalating Eritrean liberation struggle and of Somalia's drawing closer to the Soviet
Union. (28) From the late seventies, the Saudis also
developed a definite technical and economic interest in the
Red Sea. (29)

The most controversial issues in the Red Sea are those
concerning the Gulf of Aqaba and the Straits of Tiran, and
those concerning the Strait of Bab al-Mandab. In the Gulf of
Aqaba and the Straits of Tiran in the northern Red Sea,
Israel's right of passage has been established more by political factors than by legal ones: that is, through the occupation of Sinai in 1956-57, and the consequent withdrawal
which was effected in March 1957 accompanied by American
assurances for freedom of passage in return for that withdrawal. The Gulf of Aqaba is, in fact, legally a bay which
might be fairly dealt with by the littoral states - should
they so agree - as a closed bay; but this had not been possible under the existing political circumstances, and it can
therefore be regarded as territorial waters. (30)

Israel's close proximity in the northern Red Sea not
only to Egypt but also to Jordan and Saudi Arabia may, indeed, explain quite a few of Israel's sea policies, and may
well be behind both its questioning of many of the normative

provisions of the Law of the Sea during UNCLOS III and, later, its eventual rejection of the whole deal. Among other things, Israel has, at the United Nations, publicly challenged the provisions on delimitation of economic zones and the continental shelf between states with opposite or adjacent coasts. (31)

In the southern Red Sea, the Strait of Bab al-Mandab has been a source of Israeli apprehension. The narrower part of this strait lies entirely within the territorial seas of Democratic Yemen and the Republic of Djibouti, both of whom are members of the League of Arab States. Israeli sources report that during the Arab-Israeli war of October 1973, South Yemen had enforced a blockade against Israeli or Israel-bound ships. (32) Israel's vote rejecting the Law of the Sea can thus be at least partly understood in the light of the problems that surround its status and activity in the overwhelmingly "Arab" Red Sea.

On the other hand, by no means all relations in the Red Sea have been characterised by conflict. This sea has good economic potential through the utilisation of its hot brines and other metalliferous mud deposits (particularly sodium, calcium, manganese, magnesium, copper, zinc and iron). Countries bordering on the Red Sea try to coordinate their positions, and in 1972, for instance, a conference was held in which Egypt, Ethiopia, Saudi Arabia, Sudan and the Yemen Arab Republic issued a joint communiqué declaring that the deep resources of the Red Sea were the property of the states bordering it, and should remain so. (33)

Particularly significant is the fact that, in 1974, Saudi Arabia and Sudan signed an agreement that defined their respective exclusive seabed zones in the Red Sea and provided for joint exploitation of the natural resources of the area between the two zones. A Saudi-Sudanese Red Sea Joint Commission was formed and experiments conducted by this Commission show that there are reasonable deposits of various minerals in the area in which it operates. According to 1981 estimates, for example, the mineral deposits of the Atlantis II Deep revealed the existence of 2 million tons of zinc, 500 thousand tons of copper, 4 thousand tons of silver, 80 tons of gold, in addition to other quantities of lead, cadmium and cobalt. (34) Contracts are held with European and other firms for conducting research on the deposits, and factories will be set up in Yanbuc to refine these minerals. (35) Comprehensive studies on the cost of exploitation of the resources are being conducted, and the actual extraction of the resources is planned to start at the end of 1985; (36)

THE PERSIAN GULF: A SEA OF CRISIS

Even the very name of this Gulf is controversial: should it be the Persian Gulf or, alternatively, the Arabian Gulf? Or should some compromise name be devised such as the "Perseo-Arab Gulf"? (37) Iran recently threatened further action against Kuwait and other Gulf states if they did not give up the use of the term Arabian Gulf, and not long before this, the Iranian observer at a conference of radical Arab states withdrew when President Mu'ammar Qadhafi of Libya (who is quite friendly with the present regime in Iran) suggested calling the waterway the Iranian-Arab Gulf. (38)

Although several disputes in this region - such as the ones between Saudi Arabia and Bahrain, between Abu Dhabi and Qatar, and so on - have been solved, and although collective activities have been initiated - such as the signing of the 1978 Regional Convention and Protocols for Cooperation in the Protection of the Marine Environment from Pollution - by Arab countries of the Gulf, quite a few difficulties remain outstanding. These problems include the question of offshore boundary delineations between a number of countries in the Gulf region, where the availability of offshore oil resources has made them of particular significance. (39) Thus there are, for example, some problems between Saudi Arabia and Kuwait, between Iraq and Kuwait, between Saudi Arabia and Qatar, and between the United Arab Emirates and Qatar, where the difficulties have escalated to some extent recently over the Hawar island. (40) Then there are the much more difficult problems between Iran and a number of the Arab countries of the Gulf: outstanding problems with both Saudi Arabia and Kuwait over offshore boundaries, the more difficult conflict between the United Arab Emirates and Iran over the islands of Abu Musa and the Tumbs, and, of course, most tragically, the war between Iran and Iraq over - among other things - the Shatt-al-Arab estuary. (41)

Indeed, the Iraq-Iran land boundary, the offshore boundary, and the Shatt-al-Arab boundary have been subject to long historical disputes. Legally, and in practice, however,

136

Iraq has usually controlled this estuary, which represents its only outlet to the sea and which is therefore most crucial for Iraqi oil exports. As recently as summer 1982, with the fighting around the Shatt-alArab and with the refusal by Syria (a supporter of Iran) to allow Iraq's oil to flow to the Mediterranean through pipelines that extend into Syrian territory, Iraq was reduced to almost complete dependence on the pipelines to the Mediterranean that extend into Turkish territory.

From the late sixties, and at a time of rising pax iraniana over the Gulf region, Shah Muhammad Reza Pahlavi used very assertive strategies in the region. In 1969, Iran declared null and void the 1937 Irano-Iraqi Treaty on the estuary, and confronted Iraq with a show of force to back up its actions. (42) On 22 April 1969 "the Iranian freighter Ebn-i Sina, escorted by the Iranian navy and covered by an umbrella of jet fighters, negotiated the disputed waterway into the Persian Gulf... The important precedent was thus established". (43) In the meantime, Iran continued its military assistance to the Kurdish revolt in northern Iraq, carried out a military takeover of the Arab islands of Abu Musa and the two Tumbs, and sent troops to Oman to suppress the Dhofari revolution. In this atmosphere, Iraq was to sign the Algiers accord with the Shah on 13 June 1975, agreeing to delimit its river frontier according to the thalweg (median line) principle, (44) in return for the Shah stopping his military aid to and support of the Kurdish revolt in northern Iraq.

Iraq was never happy about the Algiers agreement and there was always an underlying feeling that it represented the result of a certain amount of extortion on the part of the Shah. When the Iranian revolution occurred, not only did it reaffirm all the territorial gains made in the Gulf region by the ex-Shah, but it also launched a very hostile propaganda campaign against the Iraqi regime. The Iraqi government was therefore provoked into launching a war against Iran in October 1980 in a move that is politically understandable, although legally unfounded since, among other things, the Algiers agreement had had incorporated into it a procedure for complaints made about its operation. (45)

Another sensitive issue in the Gulf relates to the Straits of Hormuz which are currently divided between Oman and Iran. The two countries have adopted the principle of "innocent passage" in administering their territorial waters, which allows them the right to restrict shipping in the Straits, a step which Iran followed in its war with Iraq and one which aroused the fears of the United States and other powers because of the fact that these straits represent an extremely "vital artery" for the oil-dependent Western economies. (46)

Arabia and the Gulf are probably the weakest point in the Arab regional system. As its name indicates, the Arabian peninsula is surrounded from almost every side by water: the Red Sea, the Arabian Sea, the Indian Ocean, and the Persian Gulf. (47) Its geo-political susceptibility, resulting from its proximity to the three major conflict areas in the region (the Arab-Israeli, the Iraq-Iran, and the Horn of Africa conflicts), combine with its large oil production and its huge oil reserves to make of the Arabian peninsula an extremely vulnerable region from a strategic point of view. (48)

"Defence of the Gulf" is therefore an issue that lends itself to frequent and controversial discussion, both within and outside the Gulf region. For example, the United States Rapid Deployment Force (RDF) is presented by the Americans to the Gulf rulers as a guarantee of their security, while many Gulfians view the RDF as, in reality, a threat to their oil resources, rather than as a protector of their national integrity. (49) The Reagan administration seems to be escalating the tones and undertones concerning American policy in the Gulf. While the "Carter Doctrine" (announced in January 1980) warned only against attempts by "any outside force to gain control of the Persian Gulf region", President Reagan went much further by stating that "... we will not permit Saudi Arabia to be an Iran". To the Gulfians, the implications are all too clear. (50)

The Arabs are not able to take such pronouncements lightly because the "interventionist" potential of American policy on the region has not been a secret. As two American naval experts indicated at the end of the seventies:

> In the event of an oil embargo which left Western Europe and Japan without adequate fuel supplies to sustain essential production, the United States is likely to mount an operation to occupy the oil fields of the Arabian Peninsula. Given the absolute dependence of Western Europe and

> Japan on oil produced in the territory of Arab
> states (i.e. given the insufficiency of non-Arab
> oil), the threat of intervention is the <u>only</u>
> deterrent to actions on the part of the oil
> producers whose consequences could be catas-
> trophic. (51)

Doubts about real American intentions may be among
the main reasons behind the paradoxical Saudi policy of
strict formal rejection of conceding military bases to
foreign powers, combined at the same time with heavy dependence on the United States in so many military matters, a
policy which has important implications for the region as a
whole.

THE ISSUE OF FOREIGN MILITARY BASES

In spite of the non-existence of any formal foreign
bases on Saudi soil, the role of the United States in "con-
structing massive bases and providing weapons and maintenance, training and advisory services at every level of the
Saudi army, air force, navy and national guard, leaves little reason to doubt that the U.S. could swiftly and effectively utilize any or all Saudi military facilities should
the situation require it". (52) Since 1965, the U.S. Army
Corps of Engineers has constructed or contracted for $12.6
billion worth of military facilities with another $9 billion
projected by 1985 on the basis of present plans. This
includes the military cities of Khamis Mushayt, al-Batin and
Tabuq, and the naval base at Jubail on the Gulf, which will
be turned over to the Saudi Royal Navy in the near future
but which will continue to be operated and maintained by a
California firm. (53)

From the point of view of the United States, there are
two targets to which American policy in the Middle East has
attached much importance in recent years: first, to "sustain
the flow of Middle East oil", and secondly, "in the case of
the U.S. overseas bases and military facilities ... to
retain our facilities as long as we can... (and)... to
provide the necessary time to develop alternatives to U.S.
global developments". (54)

In light of this, we find that although the formal
"homeporting" status of the United States in Bahrain was

terminated in 1977 (after the maintaining of a presence there since 1948), the American Department of Defense still keeps an administrative support unit in that country to carry out a number of functions that include assisting in the visits of ships and aircraft. (55)

In the same year - 1977 - the United States acquired the right to use the British territory of Diego Garcia in the Indian Ocean, a location that is conveniently small in area and sparsely populated, and it was suggested by some that this should be the base for an offshore American military facility that would take the form of a permanent Fifth Fleet, operating in the Arabian Sea. (56)

The American position in Diego Garcia is officially or unofficially supported by other states in the area that provide the United States with varying use of facilities for maintaining a Western presence in the region (such as Oman, the United Arab Emirates, Kenya, and other littoral states). The Diego Garcia facility would indeed enable the United States to sustain a policy of frequent naval visits to the Persian Gulf if the situation would so require it. (57)

As part of this strategy, efforts were made from the early seventies to woo Somalia away from the Soviet Union. Alarmed at the Soviet presence in the Somali port of Berbera, Saudi Arabia offered the Somalis a substantial amount of money in return for the "expulsion of the Russians". The Somalis duly expelled them, and then proceeded to offer the United States basing facilities at the port of Kismayu, an offer which was subsequently accepted by President Jimmy Carter. (58) By 1980, the Americans were openly seeking basing rights in Somalia, and eventually gained access for their naval and air facilities at Berbera. (59) Should the situation warrant it, the United States would also probably share France's facilities in Djibouti.

In the case of Oman, after several decades of tenure, the British ceased maintaining formal bases there in 1977, and after the revolution of 1979, the Iranians also pulled their forces out. Stepping in to replace them, the United States was thus able to gain important naval facilities in, and military ties to, Oman, especially on the island of Masirah in the Arabian Sea. Oman signed an agreement with the United States in 1980 which gives the Americans the right to use certain Omani military installations in exchange for several hundred million dollars of "aid". The country is already host to several hundred British, American and other Western military personnel. (60)

Then, of course, there are the American naval and military facilities in Egypt, both in the Mediterranean and in the Red Sea (as well as the Cairo West air base). Two joint exercises connected with the RDF, both code-named Bright Star, have already been held in the north-western desert of Egypt, and were to be followed in 1983 by a downgraded version (called Jade Green) that was to be held near the Upper Egyptian town of Qina (which was, in fact, a staging post for the ill-fated attempt to rescue the American hostages in Iran in April 1980). (61) Although Bright Star, believed to be the largest manoeuvre held by the United States in the Middle East since the fifties, was conducted mainly in Egypt, other related aspects of the exercise are also believed to have been held simultaneously in Oman, Somalia and the Sudan.

Since the Israeli invasion of Lebanon, which took place in June 1982 with apparent American condonation, there seems to be some reluctance on the part of the Egyptian authorities to allow the United States to develop fully the Red Sea port of Ras Banas, as a base that would be connected with the American RDF. In the meantime, the controversy continues within many Arab circles as to whether the main target of the RDF is the Soviet Union and its large encroaching forces, or whether it is really the Arab peoples and their rich oil fields.

CONCLUSION

One can conclude in general by saying that the Arab states have, on the whole, accepted the main orientations of what may be termed customary international law concerning the sea. Their attitude in the recent conferences that resulted in the adoption of the new package deal over the law of the sea has largely reflected variations on the general "Third World approach". (62) Their positions derive mainly from socio-economic rather than from military-strategic considerations. Thus, for example, reference was made to such economic considerations in the continental shelf proclamations issued at the end of the forties by the Gulf states, and then subsequently by other Middle Eastern countries. The important claim by Saudi Arabia in 1949 defining its territorial waters was not the result of immediate security interests, but was indirectly brought about by economic motivations and a desire to provide for off-shore oil exploration and exploitation. (63) The extension to twelve miles of the territorial sea was based essentially on economic criteria, which also provided the reason for the exclusive fishing zones recently claimed by certain Gulf states.

The rather "standard" attitudes of most Arab countries do not by themselves, however, guarantee the end of all sea conflicts in the Middle Eastern region, since its countries tend to have different economic and political interests. Thus, for example, the status of Israel in the Red Sea area remains quite sensitive, and some fairly bloody sea conflicts do actually occur, the most notable - as we have seen - being that between Libya and the United States over the Gulf of Sidra in 1981, and those between Iraq and Iran over the Shatt-al-Arab estuary since 1980.

The strategic vulnerability of the Arabian peninsula has been a subject for great concern in many quarters in recent years. But while Arab countries of the Gulf (with the possible exception of Oman) have been trying to maintain a principle of at least no formal concession of naval or military bases to foreign powers, the United States is still seeking more facilities with "friendly" regimes and is, at the same time, developing its controversial Rapid Deployment Force, which many analysts believe is meant largely for possible utilisation in the Gulf region.

Finally, considering the region as a whole, and keeping in mind the future potentials, there is still a great deal that can be done towards increased functional cooperation, in areas such as scientific research, pollution control, joint ventures and the like, and one can only hope that the existing political and legal disagreements between certain Middle Eastern states, as well as the superpower rivalries involving the region, will not hinder too severely the prospects for mutually beneficial collective endeavours.

NOTES

1. Robert O. Keohane and J.S. Nye, Power and Independence: World Politics in Transition, Little, Brown, Boston, 1977, pp. 97, 121-126, 148-150.

2. Ann L. Hollick, "The Third United Nations Conference on the Law of the Sea: Caracas Review", in Ryan C. Amacha and Richard J. Sweeney (eds.), The Law of the Sea: U.S. Interests and Alternatives, American Enterprise Institute, Washington D.C., 1976, pp. 126-127.

3. Cf. Saudi Report, 7 June 1982, p. 8.

4. Bernard H. Oxman, "The Third United Nations Conference on the Law of the Sea: the Tenth Session, 1981", in American Journal of International Law, Vol. 76, No. 1, January 1982, pp. 10-11.

5. Ibid., p. 6.

6. There were 130 votes for the package, 4 against, and 17 abstentions. Turkey, Israel and Venezuela joined the United States in opposing the treaty. The Soviet bloc (with the exception of Romania) abstained along with Belgium, Britain, Italy, Luxembourg, the Netherlands, Spain, Thailand and West Germany. France, Canada and Japan were the main "Western" countries voting with the majority. The Guardian, 9 May 1982. As one distinguished editorial put it, if the new code was to founder, the seas would probably not disintegrate into anarchy, but a unique opportunity to bring more order into the world's marine affairs would have been lost. The Economist, 17 July 1982. Fortunately the convention was signed by 118 countries on 10 December 1982, and other countries may have up to two years to do so. Los Angeles Times, 11 December 1982.

7. Oxman, op. cit., p. 20.

8. Ali A. El-Hakim, The Middle Eastern States and the Law of the Sea, Manchester University Press, Manchester, 1979, p. 42.

9. Ibid., p. 46.

10. Ibid., p. 78.

11. Charles G. MacDonald, Iran, Saudi Arabia and the Law of the Sea: Political and Legal Development in the Persian Gulf, Greenwood Press, Westport, Conn., 1980, p. 170.

12. El-Hakim, op. cit., pp. 60-79.

13. Ibid., p. 9.

14. Los Angeles Times, 22 August 1981.

15. Los Angeles Times, 23 August 1981.

16. Los Angeles Times, 4 September 1981.

17. El-Hakim, op. cit., p. 10.

18. Ibid., p. 215.

19. Middle East Journal, Vol. 36, No. 1 (Winter), 1982, p. 79.

20. Ibid.

21. Ibid., pp. 79-80, and refs. quoted therein.

22. Keohane and Nye maintain that superpowers are nowadays declining to use force over sea issues, which increases the manoeuvrability of small states. Keohane and Nye, op. cit., p. 126. However, as the Third UNCLOS was convening, an American professor of Marine Law was suggesting, after angry condemnation of OPEC and the ruling elites of Third World countries that: "With respect to virtually all of the issues involved, the use of force is a possible method for pursuing U.S. security and economic objectives where these objectives conflict with the unilaterally or multilaterally established regimes of coastal developing nations. Force could, for instance, be used to secure access to 200-mile fishing zones of other nations for distant-water fishermen on the basis that waters beyond twelve miles were subject to high seas freedom of fishing; force could be used to protect deep-seabed mining operations being conducted contrary to a Group of 77 seabed treaty; or force could be used to protect merchant shipping in economic zones, territorial waters, or straits... " H. Gary Knight, "Alternatives to a Law of the Sea Treaty", in Amacha and Sweeney, (eds.), op. cit., pp. 141-142.

23. Los Angeles Times, and Al-Mustaqbal, June-November 1981.

24. El-Hakim, op. cit., p. 34.

25. Ibid., p. 24.

26. Middle East Economic Survey, Vol. XXV, No. 51, 4 October 1982.

27. Cf. Abdullah A. Al-Sultan, "The Arab-Israeli Interaction in the Red Sea: The Implications of Two Contending Strategies", (unpublished Ph.D. Dissertation), University of North Carolina, Chapel Hill, 1980.

28. Colin Legum, "The Middle East and the Horn of Africa: International Politics in the Red Sea", in Colin Legum and Haim Shaked (eds.), Middle East Contemporary Survey, Holmes and Meier, London, 1978, pp. 60-61.

29. A sea charting centre was established in Jeddah

recently (cost $8.9 million) to map the Red Sea and, eventually, the Arabian Gulf. Saudi Report, 9 August 1982.

30. El-Hakim, op. cit., pp. 167-177.

31. Oxman, op. cit., p. 14.

32. Cf. Mordechai Abir, Oil, Power and Politics: Conflict in Arabia, the Red Sea and the Gulf, Frank Cass, London, 1974, pp. 200-201; and his Persian Gulf Oil in Middle East and International Conflicts, The Hebrew University, Jerusalem, 1976, pp. 20-21.

33. El-Hakim, op. cit., pp. 178-188.

34. The Saudi-Sudanese Red Sea Joint Commission, 1981-82.

35. The town of Yanbuc on the Red Sea is being developed as a major port city, with a planned population of 150,000. Saudi Report, 27 September 1982. It is expected that Yanbu^c will be useful in the export of crude and gas in the event of any blockade in the Persian Gulf. Oman has also proposed the extension of an oil pipeline to pass through its territory and by-pass the Straits of Hormuz. Al- Mustaqbal, Vol. 5, No. 239, 19 September 1981.

36. Saudi Arabia Yearbook, 1980-81, pp. 62-63.

37. Cf. Sayed H. Amin, International and Legal Problems of the Gulf, Menas Press, London, 1981, esp. Chapter 2.

38. Cf. Nazih Ayubi, "Arab Relations in the Gulf", in S. Tahir-Kheli et al. (eds.), The Iran-Iraq War: Old Conflicts, New Weapons, (Praeger, New York, 1983).

39. MacDonald, op. cit., pp. 33-36.

40. Cf. Al-Mustaqbal, 3 April 1982, pp. 32-35.

41. Cf. Amin, op. cit., Chapter 4.

42. MacDonald, op. cit., p. 54.

43. R. K. Ramazani, The Persian Gulf: Iran's Role, Virginia University Press, Charlottesville, 1972, pp. 43-44.

44. MacDonald, op. cit., p. 34.

45. Richard A. Falk, "International Law and the Peaceful Settlement of the Iraq-Iran Conflict", in A. Dessouki (ed.), The Iraq-Iran War: Issues of Conflict and Prospects for Settlement (Policy Memorandum No. 40), Center for International Studies, Princeton University, 1981, p. 80 ff.

46. MacDonald, op. cit., P. 5.

47. In recent years countries of the Arabian peninsula have been showing signs of trying to make good use of their location. Among other things, there is a definite interest in expanding and improving shipping activities (including ship-building) being shown by Saudi Arabia, Kuwait and the United Arab Emirates. Saudi Report, 7 June 1982; 21 June

1982.

48. Even Saudi Arabia, the hegemon of the region, is quite poor in military capability. In 1981 the Kingdom had no more than 51 thousand active military personnel (and 20 to 30 thousand members of the National Guard), supported by 139 combat aircraft and 636 tanks. The navy capability — which is of particular relevance to this essay — remains quite restricted, in spite of the country's long coastline. With a mere 1,500 men, inadequate equipment and limited patrol ability, the Saudi naval forces are likely to be insufficient for the country's maritime defence, even on completion of the current modernisation programme. The air force is probably the most advanced branch of the armed forces: although it has only 14,500 personnel, its equipment is very advanced and includes five AWACS surveillance aircraft (which, however, may have to be maintained and operated almost indefinitely by American personnel). Cf. International Institute for Strategic Studies, The Military Balance, London, 1981, Armed forces Journal International, September 1981, pp. 56, 61, 78-80. For further details and analysis see also: Nazih Ayubi, "Vulnerability of the Rich: the Political Economy of Defence and Development in the Gulf"', Center for Strategic and International Studies, Georgetown University, Washington D.C., June 1982.

49. Christopher van Hollen, "Don't Engulf the Gulf", Foreign Affairs, Summer 1981, pp. 1066-1068; Abdul Kasim Mansur, "The Military Balance in the Persian Gulf: Who Will Guard the Gulf States from their Guardians?", Armed Forces Journal International, November 1980, pp. 48 ff.

50. Joseph T. Malone, "Saudi Arabia: The Pace of Growth and Spreading Power", Middle East Problem Paper No. 17, The Middle East Institute, Washington D.C., 1978, p. 7; Valerie Yorke, "Security in the Gulf: A Strategy of Pre- emption", The World Today, No. 6, July 1980, p. 244.

51. Edward N. Luttwak, Sea Power in the Mediterranean : Political Utility and Military Constraints, The Washington Papers, Vol. VI, No. 61, Sage Publications for CSIS, Georgetown University, Beverley Hills, 1979, p. 48.

52. Joe Stork, "The Carter Doctrine and U.S. Bases in the Middle East", MERIP Reports, No. 90, September 1980, p. 10.

53. The United States' military-related presence in Saudi Arabia is estimated at well over 10,000 personnel. Stork, Ibid., and refs. quoted.

54. Alvin J. Cottrell and T.H. Moorer, <u>U.S. Overseas Bases: Problems of Projecting American Military Power Abroad</u>, The Washington Papers, Vol. V, No. 47, Sage Publications for CSIS, Georgetown University, Beverly Hills, 1977, p. 57.

55. Stork, <u>op. cit.</u>, p. 11.

56. James R. Kurth, "American Leadership, the Atlantic Alliance and the Middle East Crisis", in Steven L. Spiegel (ed.), <u>The Middle East and The Western Alliance</u>, George Allen and Unwin, London, 1982.

57. Cottrell and Moorer, <u>op. cit.</u>, pp. 58-59.

58. Fred Halliday, <u>Soviet Policy in the Arc of Crisis</u>, Institute for Policy Studies, Washington D.C., 1981, pp. 102-103.

59. <u>Ibid.</u>, p. 105.

60. <u>The Economist</u>, 20 November 1982, p. 53.

61. Anthony McDermott, "Egypt and the U.S.", <u>Middle East International</u>, No. 186, 29 October 1982, p. 6.

62. El-Hakim, <u>op. cit.</u>, p. 191.

63. MacDonald, <u>op. cit.</u>, pp. 104-105.

Part Three

ASPECTS OF POLITICAL AND MILITARY CONFLICT

Chapter Five

MAGHREBI POLITICS AND MEDITERRANEAN IMPLICATIONS

I. William Zartmann

INTRODUCTION

Neither divine will nor geographic determinism nor any other permanent destiny has imposed fixed and immutable causes of conflict (or cooperation) in this world. Objective factors are subject to actors' perception, and subjective factors are highly variable. This is not to say that events are haphazard and without cause, but simply that even the most longstanding quarrels and friendships are the outcome of different ingredients, interests, desires and decisions from regime to regime, country to country, era to era. To identify ongoing sources of conflict in North Africa, then, is an exercise in interpretation rather than a study which identifies permanent, overriding factors within which rulers try to manoeuver. (1)

Furthermore, in North Africa, many of the strongest sources of conflict operating in other areas are absent. The four Maghrebi societies discussed here - Morocco, Algeria, Tunisia, and Libya - are brothers. All are Arabo-Berber societies, practicing Sunni Islam, speaking mutually intelligible if slightly variant dialects of Arabic and reading a common newspaper standard (with a few not-yet-Arabised Berber pockets), and following generally the same lifestyle. All underwent the same conquest and Arabisation process and evidenced the same successful assimilation practices that might be the envy of later-day colonialists. When the Muslim community fell under Turkish rule, the Sherifian Empire (and the Saharan nomads who owed it spiritual allegiance) escaped its suzerainty, without becoming a historic enemy to the beyliks of Algiers, Tunis or Tripoli. When the European conquest came, over the century after

1830, three countries fell under the same French ruler, then
became independent through the efforts of similar and
related nationalist movements, between 1956 and 1962. Ita -
ly's brief conquest of Libya ran from 1911 to 1942.

The societies which continued their collective existence through these events show greater differences within
than among themselves. On a map without boundaries, there is
nothing to indicate where any of the component states starts
and stops. Instead, one sees roughly east-west bands of features cutting across the region, wider or higher in the
west, pressed into the sea by a north-pushing desert in the
east. The predominant feature is a mountain chain that
starts in the Canary islands and cuts diagonally across
Morocco, joined by other spurs from the Atlantic coast and
the Mediterranean shore, flattening out into the High
Plateaus of Algeria and the Tell of Tunisia before dipping
into the sea and emerging as Sicily and the Appenines.
South of the Atlas chain, the land is desert and valuable
for its mineral resources alone. On the highlands and
plains, the land is good for farming, and a number of rivers
rise from the mountains to provide current irrigation, with
a potential for expansion. The vocation is not new: North
Africa fed Rome in an earlier time. The population has
molded itself to these geographic bands. It is densest along
the plains and thins out into the interior, although this
distribution was not always so. It was European colonisation
which turned economic activities towards foreign trade and
established or expanded the large port cities; the earlier
Arab conquest flowed westward in the interior of the country
and located its cities on land trade routes. The flood of
Arabisation left significant pockets of Berbers above its
high water mark - most in Morocco (35%) where the mountains
are highest and most inhospitable, a smaller group (18%) in
the Algerian Kabylia, least (5%) in less mountainous
Tunisia. Bands of social similarity run across the region,
so that there is more in common among the lifestyles of the
mountain herdsmen of the Atlas, the Kabylia, Jebel Akhdar,
the Khroumeria, of the plainsmen of the Gharb, the Mitidja,
the Mejerda, or of the traditional artisans of the medinas
of Fes, Constantine, Tripoli, and Tunis, than there is among
the several levels within each country. The same effect is
even more characteristic of relations between the modern
sector of businessmen, military, bureaucrats, technicians,
and intellectuals and the more traditional parts of each

country.

Many more characteristics could be cited to show similarities and diversities in North Africa, and other elements will be brought out as specific causes of conflict are identified. But the major point from which to begin is a recognition that great cultural clashes, like the Hindu-Muslim conflict on the Indian subcontinent, or longstanding national drives, like the Russians' search for warm-water ports, or historic conflicts reinforced by national character interpretations, like the politics of the French and the Germans or of the French and the British, or large-scale geopolitical imperatives, such as the various interpretations of heartland and rimland, or geo-historical patterns, such as the ebbs and flows between Nile and Fertile Crescent across the landbridge of Palestine, have no equivalent in the sources of conflict in the Maghreb. Conflict in North Africa is political, not social or historical, and is even functional to the development of the state.

GENERAL SOURCES OF CONFLICT IN MAGHREBI POLITICS

Borders

The most obvious potential source of conflict between any states is found in their boundaries. (2) Since North African states contain nations abuilding and since the concept of a territorial bounded state is relatively new to the region, boundary problems are to be expected. To this should be added the fact that the terrain is often hard to mark and was long considered not worth marking. Under such conditions, it is not hard to understand why there simply is no established boundary in some areas.

The Muslim socio-political unit is the community of believers (umma), ruled by a representative (khalifa) of God chosen by the community and in contractual (bei'a) relationship with it. This "state" existed wherever its members moved, resulting in a demographic rather than a territorial unit. Obviously, no state can be exclusively one or the other but it is the basis of the concept that is referred to here. Since the jurisdiction of the state was determined by the allegiance of the people, not by the limits of the land on which they lived, there were no fewer "territorial" problems but there were no boundary problems per se. Where nomads settled, city-states and their regions gave their allegiance to one capital or another but the limits where one region abutted on another were usually not under any authority strong enough to require a line of demarcation. Even in the era of pre-modern national consolidation, at the end of the eighteenth century (Bey Hamuda II in Tunis, Dey Mohammed ben 'Uthman in Algiers, Sultan Mawlay Mohammed III in Fes, Yusuf Karamanli in Tripoli), it was the central authority which was being consolidated but not the geographically finite extent of its writ.

Under French conquest, some boundaries were established to separate colonised (Algerian) from not-yet-colonised (Moroccan and Tunisian/Libyan) territories; when colonisation was completed, any disputes were at least an internal, French affair. A large part of the territory was treated in the same way as the North African rulers themselves had done earlier, as a region to be shifted from one jurisdiction (Algiers) to another (Rabat) but unworthy of an established boundary since it was "without water" and "uninhabitable" and so a boundary would be "superfluous". In the absence of any such boundary determination, the way was open after independence for nationalist movements to revert to a traditionalist justification for irredentism. Particularly where the colonial rule took the form of a protectorate of an ongoing monarchy and khalifate, as in Morocco, where in addition only the main part of the realm received its independence at first and bits and pieces were restored thereafter, it was logical to push for a maximum restoration of the Empire. On the other hand, where the restoration of the monarchy was not the basis of nationalism but where an undemarcated frontier ran into a desert that was considered to be a common patrimony, it was logical to call for a redrawing of the boundaries, and then equally logical to be satisfied with a solution that provided for negotiated sharing of the benefits of the desert.

The traditional basis of border disputes could be expected to pass away along with other aspects of traditional legitimisation, were it not supported by two other, more modern concerns. One is the discovery of mineral deposits. The Algero-Libyan dispute was introduced by some old maps and unratified treaties but is exacerbated by oil under the ground and is still not settled in the minds of the current Libyan rulers. Twenty years earlier, the Algero-Tunisian dispute was ignited by a Tunisian agreement to run a French pipeline from the Edjele oil deposits to the nearest port, the Tunisian city of Skirra, during the Algerian war when the question of sovereignty over the Sahara had not yet been settled; the exacerbant of the conflict was the discovery of an oilfield at el-Borma initially tapped from the Algerian side but with most of its underground deposits on the Tunisian side of the border. The Algero-Moroccan dispute was compounded by the presence of an iron mountain at Gara Jebilet, south of Tindouf, which the Moroccans claimed and which the Algerians could exploit most economically for export through a Moroccan port. It was further complicated by the discovery of rich phosphate deposits at Bou Kra' in the northern part of the then-Spanish Sahara (in this part of the Moroccan irredentist conflict, the question is not over boundaries but ownership).

The other "modern" element is the role of success in legitimising the claim and claimant. In each dispute, the national leader of each country staked a lot on his claim, and only Bourguiba has been secure enough in other aspects of his program and legitimisation to be able to withdraw and compromise gracefully. The bitter dispute over the Sahara has continued beyond all reason, with an Algeria which has no direct claim on the territory challenging a relatively well-established fait accompli by Morocco; yet the endurance of the conflict is somewhat more understandable when the personal engagement of the two leaders, both in the process of a major election and a shift in political institution-alisation, is considered. By the same token, the fact that both have come through their domestic political challenge in good shape may permit some disengagement on the border issue.

The resolution of the contemporary boundary disputes in North Africa has displayed some very specific character-istics. There have been no disputes over established, de-marcated boundaries other than the Libyan special cases, and most of the disputes have concerned the disposition of colonial territory. In addition, those disputes which went to war followed the usual African pattern of brief hos-tilities until the exhaustion of current stocks, followed by a cease-fire and then a long period marked by bits of progress and regress in the resolution of the conflicting claims themselves. The Fabian proxy war in the Moroccan Sahara is a new military approach to contemporary conflict but it resembles traditional means and even traditional behaviour in general among the nomads (particularly the

153

Reguibat) of the region. Even when boundaries are established and demarcated, frontier mineral deposits, transit facilities, and nomadic wanderings will continue to provide the
basis for conflict, although at a much more manageable level.

National Consolidation Processes

If boundary disputes refer to the shell of the state,
national consolidation as a source of conflict refers to the
internal composition within that shell. (3) The political
history of North Africa over the past two centuries (or
more) can be interpreted as a struggle for national consolidation, an attempt to create a centralised governmental
structure capable of commanding the loyalties and mobilising
the resources necessary to meet the demands of modern life.
The need to build a centre implies also the need to define,
attract, and hold its periphery, and this is often a competitive process with one's neighbours, both because of
potential conflict over the same peripheral territories and
because patriotic solidarity against a consolidating neighbour is often the best means of consolidation at home. Some
aspects of this process have already been noted in connection with boundary disputes, for there is an inevitable
overlap.

Of the major consolidators at the end of the eighteenth century, only the Moroccan sultan and the Tunisian bey
were successful to any extent, and the bey - favoured by
geography and history - was the only one who could boast a
centralised state with the beginnings of a bureaucracy and a
sense of national identity, organised about a single
capital, and enough in control of its functions to be able
to experiment not only with military and tax organisations
but also with industry and education. Morocco was in a sense
a Tunisia spread out over a larger territory, in which the
sultan occupied the major cities as his capitals and
attempted to consolidate his temporal control over the rest
of the territory; the effort required left little energy or
structure for the activities of modernisation that Tunis was
able to begin, even if unsuccessfully. In Algeria and Libya,
the writ of the city-state authorities did not run far
beyond their walls and the centralising power was absent. As
a result, when the French came to conquer a coastal enclave
- as they and the Iberians had been doing for the past three
centuries - they found themselves drawn both by vacuum and
by dissidence into the interior, and eventually into contest

154

with the Sherifian Empire over whole border regions. But more important is the fact that even before the French arrived, the sultan sought to include the region of the Tlemcen city-state in his consolidating realm, and in the process of fighting the French, he sought to extend the same suzerainty over Abdelqadar. In Libya, the Italians did nothing at all to consolidate a state.

The territorial division of the Sahara can be seen as part of the same conflict of national consolidation and contests over the periphery, a process of sorting out national territories, identities, and activities in which the centre tries to consolidate its ability to control its land and people and at the same time tries to establish convincing reasons why a particular territory and population should be part of its state rather than of another. Inevitably, there is a conflict potential inherent in both types of activities as states attempt to protect themselves against and distinguish themselves from their neighbours. The constant parallel between Morocco and Algeria in such military measures as arms imports and military budgets and the sudden overarming of Libya and then of Tunisia are indicators of the first. A less frequently cited example of the second is the way in which one state uses the other as a target of nationalism: Ben Bella used the Algero-Moroccan war of 1963 to rally the Kabyle dissidents to support his regime, Bourguiba used the Gafsa raid of 1980 to reintegrate the opposition, and Hassan II used the entire irredentist issue, but particularly the Green March into Spanish Sahara in 1975, to swing over the entire opposition to his camp. In the case of unconvertible opponents – whether they be Morocco's ben Barka, Algeria's Zbiri and Qaid, or a series of Tunisian plotters and politicians – one of the components of their treason is their support from the neighbouring country, even though no state of war exists between the two. There is of course an ideological dimension that goes beyond simple national consolidation and that is associated with this cause of conflict, but it will be the subject of separate examination in the following section.

Before turning to ideology, however, there is an additional aspect to national consolidation that warrants discussion. The attempts of the North African states to solidify national structures and develop national identities is conflictual because the societies – and indeed the economies – are so similar. If the countries were visibly different, national consolidation would not require such effort; if the economies were complementary, cooperation would be easier. Too much can be made of this point: Maghrebi states are not perpetually at loggerheads because they grow Mediterranean products, but, although growing Mediterranean products does give them certain common interests, it also provides the necessity for working out shares and competing for markets in dealing with the outside world since precolonial times.

The reverse is more significant: Since Maghrebi states are going through a process of national consolidation, protectionist autarky often takes the lead over interdependent complementarity. Complementarity is seen as dependency, subject to the whims of capricious leaders of rival states. It is hard to find pre-colonial, pre-industrial examples of Libya's, Morocco's, Algeria's and Tunisia's reluctance to locate an all-Maghrebi steel mill in one of their countries; heightened self-sufficiency seems to come with modernization. The conflict appears with a neo-Leninist extension of this activity: Given the size of the North African states' internal markets, the drive for self-sufficiency paradoxically requires the conquest of a regional market to be successful. Thus, Algerian industrialisation is predicated on Algerian dominance of the Maghrebi market, a policy that cannot work if each state adopts it.

National social, political and economic consolidation is an imperative felt by each of the North African states, pressing them into distinctions, competition, and imposition. Conflict here is more subtle than in the case of boundaries, and it is hard to distinguish particular patterns to it. Nor is the obvious solution offered by comparative advantage - in economics or in other areas - always an acceptable means of conflict resolution.

Ideology
    A third source of conflict is ideology, which adds a further dimension to the previous factors. (4) Polities are not merely interchangeable entities, behaving identically under the same imperatives of external and internal consolidation as modified only by free will and accidents of geography. States also compose the myths and values by which they live, and these ingredients of ideology are indissociable from the structure of society and its political reflection. The North African states have identified and reinforced their objective differences with ideological per-

ceptions, colouring a black-and-white picture so as to bring out some aspects and hide others. Such ideological perceptions tend to be self-proving hypotheses, providing a subjective basis for conflict that often outlasts the original objective factors. The perceptions thus appear to have a longer future than a past; although they can be consciously altered, they seem to be a feature of developing, industrialising society, with no precolonial antecedents.

Algeria is basically a revolutionary proletarian society. It was twice revolutionised, by the French when they came and by the French when they left, independent of anything the Algerians themselves might do: The first move superimposed an ascriptive upper class on society, the second decapitated the social structure, and the Algerian response of national solidarity and social promotion was a logical reaction. In a country where few of the new ruling elite owned land, where any remaining aristocrats were collaborators, where the national liberation struggle was a people's (and even a peasants') war, and where cultural identity had been kept alive in the home, the political myths of the new state were inescapable.

In Libya, the first independence was a traditional restoration, with a small bourgeoisie making money under the protection of the palace (itself inhabited by an old sufi symbol), but the second independence (to borrow an African opposition term) was a traditional Jacobin revolution, with ideas for change imposed from the top but drawing their inspiration from a number of sources within the society (all parallels break down somewhere, and unlike the original Jacobins the Qadhafist ideology drew heavily on religion for its ideas). Earlier social and political institutions − bourgeoisie and palace − were new and weak in legitimacy, inviting an ideology in their mirror image.

In Morocco, by contrast, nearly every factor was different and a bourgeois monarchy was the result. Both the commercial aristocracy and the tribal monarchy were protected by the Europeans, even before the formal French Protectorate which strengthened both and then gave way to their nationalist alliance. Both were landowners, both collaborators, both nationalists, and both maintained the cultural identity during the two-generation-long French occupation. Even when the king has tried to pursue a radical foreign policy, as in the Casablanca Group of 1961-62, the words never sounded the same as when used by the Algerians.

157

Maghrebi Politics and Mediterranean Implications

In Tunisia, the polity has been a bourgeois republic, with the baldi families of Tunis playing a leading role in the governance of society even under the bey and the bourgeoisie of the lesser cities of the Sahel joining them through the nationalist movement. Conveniently, it was the bey who was the collaborator, but property ownership, cultural identity, and political participation belonged to the rest of the population. Political party activity, a government exercise in Algeria and a (loyal) opposition exercise in Morocco, was typically an integrated exercise of rulers and ruled in Tunisia. These elements may be impressionistically selected but they are facts nonetheless, and they form the basis of very different ideologies. They are subject to both evolutionary and revolutionary change, to be sure, but such changes can also be identified nationally.

The result is a number of empirical referents for self-perception and identification that influence political discourse. Societies not only see themselves in chosen terms but they see others in relation to these terms, either applying their referents to other societies or using or contesting the terms other societies have chosen. Colonial farmers generally agreed that Moroccan farm labourers were better than Algerians because they were harder working and accepted authority. Moroccan students in the 1960s longed for the socialist intellectuals' paradise which they perceived Algeria to be. Algerians felt assured that Morocco's feudal monarchy would collapse in its own corruption. In the 1970s, readers of Moroccan newspapers and students in the University had to be satisfied with biased and incomplete information on the Algerian system. On the basis of their own and others' ideological perceptions, similar and neighbouring societies have begun to make real distinctions between each other, building on false expectations and poor information, and with a mixture of envy and disdain seeing the enemy of the domestic system in control of the system next door.

The thesis presented here is that such artificial distinctions which make themselves real are indeed the reflection of the historical evolution of societies but are also the product of a felt need to make oneself identifiable and distinct from one's neighbours where otherwise such distinctions do not exist. If there were no real ideological differences, they would have to be invented. The continued ideological rivalries of Egypt, Syria, and Iraq, or of Russia and China, are applicable to the Maghreb, not inapplicable because of North African internal similarities. A revolution in Morocco will not bring long lasting amity with Algeria, for reasons of both reality and perception. In fact, only when identities are sorted out and given a real distinction internally and acceptance by others externally will ideology receed as a source of conflict.

The Checkerboard Pattern of Relations

A fourth source of regional conflict, again related to the previous factors, is the checkerboard pattern of relations characteristic of North Africa. (5) According to this basic configuration of international relations, the states of the region have continually acted as if "my neighbour is my enemy and my neighbour's neighbour is my friend". Four countries, standing side by side along the Mediterranean coast, with their backs to the sand and their faces to the sea, held together by a common notion of regionality, by membership in the same overlapping regional international organisations (the Organisation of African Unity and the League of Arab States), by an eastwest transportation network, and by the need to interact, have had to develop a pattern to these interactions. In addition, this pattern was encouraged by the French during the nationalist period and thereafter to the present. Even the institutionalised attempts at regional unity have been dominated and broken by this same pattern.

The very act of granting independence to Morocco and Tunisia in 1956 was an attempt by France to secure its Algerian flanks, cut its losses, and enlist new allies in its North African policy. While these aims failed, the attempt of the three nationalist movements to establish the basis of regional unity the following year ran ashore on the effect created by the French, namely that two of the nationalist movements were government parties while the third was a rebel guerrilla force. Later that year, Tunisia persuaded the Algerian FLN to move its headquarters from Egypt to Tunis, and alienated the Sherifian Empire by abolishing its own monarchy. Tunisia then alienated the FLN by allowing the Ejele-Skhirra pipeline to cross its territory, and in 1961 Morocco had enlisted it in its first alliance, the Casablanca Group, from which Tunisia was excluded because of its support for Mauritania against Morocco. Within two years, Algeria was independent and the two former allies were at war with each other.

Beginning in 1965 with the Hassan-ben Bella meeting at Saida (which in part resulted in the latter's overthrow four months later), the pattern began to change slightly. Algeria's Maghrebi policy has been to develop a pattern of friendly bilateral relations with its neighbours as an alternative preferable to predetermined hostilities and more realistic than regional unification. Boumedienne worked out agreements with Morocco over trade, mineral exploitation, foreign policy, and finally the border, in 1968-72, settled the el-Borma dispute with Tunisia in 1970, joined in positive-sum tripartite cooperation with Morocco and Mauritania over Spanish Sahara against Spain, and signed a military alliance with Libya in 1975. None of it lasted, except perhaps the Libyan alliance. Relations with Tunisia have become testy on occasion, and relations with Morocco have moved back to the edge of war since 1975.

Algeria's policy is well-fitted to its position: It is the keystone state, and any move towards regional unity is impossible without or against it. But Algeria is not interested in regional unity either, since such an arrangement would saddle it with more problems than advantages. As a result, a position of policy dominance is all it seeks to achieve, but the tendency to return to a checkerboard pattern of relations has proven irresistible.

Into this checkerboard pattern then came Libya, essentially after its Mashreqi rebuff in the October war of 1973. After its abortive unification with Tunisia in 1974, it set up the basic pattern of relations for the coming decade by signing the mutual defence treaty of Hassi Messaoud in 1975. The two allies sometimes act as enemies (over the border), sometimes as rivals (over Saharan, Chadian or Tunisian policy), sometimes even as allies (in the Steadfastness Front) - as a checkerboard set of neighbours/non-neighbours should.

Except for a short time in the eleventh century, North Africa has never been a political or economic unit. All colonial rulers divided the extensive territory into administrative units with different regimes. There is no concept of a "Maghrebi nation" as there is of an "Arab nation", despite the interest in Maghreb unity. In fact, there is not even a single Arabic word for North Africa, since "Maghreb" refers properly to Morocco. Nor is there a common enemy on the horizon, against which the states could unite.

Whether the checkerboard pattern is played out at the level of hostilities or merely as shifts from disillusion to indifference, it appears to be a resiliant tendency, perhaps more a reflection of relations than their cause, but nonetheless a pattern that seems to reassert itself despite some efforts to the contrary.

Position Within Regional Contexts

All of the previous sources of conflict have been essentially internal to the states or to the region. A fifth source concerns the states' relation to the two larger regions to which they belong - Africa and the Middle East. (6) For long the states of Africa have had a fair degree of equality in terms of effective power resources. As a result, despite some early attempts, no continental leaders have arisen; no states have had the resources to support a bid for African leadership and even the best-endowed states have not been able to mobilise and transform these resources into power. The same characteristic is obviously not true of the Middle East, where leadership struggles have been endemic, but it is applicable to North African participation in these struggles.

However, this situation is changing, particularly in regard to the African states. Some states in Africa - among them the three states of North Africa - are developing at a respectable rate, pulling far ahead of others, and this development can be translated into applicable power and into the desire to exercise power. As a result, the coming decade is likely to see serious leadership struggles, particularly in regard to Africa, with the Maghrebi states, particularly Morocco and Algeria, drawn in in pursuit of their potential.

Morocco and Algeria are twins, as already noted in part, a condition which accentuates their rivlary. With about 20 million people each, they are the sixth and seventh largest states in Africa (including Egypt). With about $16 million and $32 million GNP, respectively, they are the third and sixth richest states in Africa; Libya with $25 million is fourth. With percapita GNP of $900 and $1800 they are still the fourth and sixth richest states (Libya with $9,000 and Tunisia with $760 are first and fifth by this measure); their per capita growth rates over the 1970s decade of 3.9 and 2.6% show a steady growth performance, but Tunisia's per capita growth rate of 5.8% in the 1970s is outstanding. Libya, over the mercurial decade, fared badly

161

at -2.6%. Tunisia and Libya have universal primary education and Algeria may attain it by the 1980s; Algeria and Morocco can be expected to have 20% of the eligible-age children in secondary school in the 1980s, with Tunisia 30% and Libya 50%, and figures of 3% to 5% for higher education, giving them a sizeable pool of skilled manpower.

A few other factors work against Moroccan and Algerian pretentions at African leadership, particularly their distance and separation from the rest of the continent and their racial difference from Black Africa.   On the other hand, their rivals are few: The obvious one is Nigeria, if it rouses itself, but the more proximate rival is Libya even though its pretentions are primarily limited to Arab Africa and may not outlast the current regime.

As   North   African states develop ahead of others in their region, they will be increasingly tempted to use their sources of power to line up other states behind them and save the rest from the misleading influences of their rivals. Leadership is therefore not likely to be satisfied with its followers of the moment, it is also likely to seize on the developmental problems of its rivals' followers to attack and subvert their regimes if it cannot influence their decisions.

It would be presumptious to seek historic roots for such policies in the eleventh-century Saharan Empire of the Almoravids   or the sixteenth-century trans-Saharan expeditions of the Saadis, or in the wide-ranging colonial alliances and conquests of the Italians and French. The most immediate antecedents come from the incipient balance of power politics that arose out of the two groups of new African states, named after their founding capitals, Casablanca and Brazzaville, in 1959-1963, prior to the creation of th OAU. Subversion was a means of carrying out inter-African relations during this period, in the absence of a conflict-resolving forum and norms provided subsequently by the African Organization. The establishment of the OAU generalised African states' contacts with each other and their concern over African problems, while at the same time creating a concert pattern of international relatoins that kept blocs and alliances from forming.

The rise of more developed states and the breakdown of the conflict resolution machinery of the OAU concert system has brought back the predominance of camps and blocs, and the sides have hardened. The trigger has been the Saharan affair, which deadlocked the OAU into two equal groups behind Morocco and Algeria; it could also be said that the Angolan affair provided the initial split, but the intervention of South African undermined the "moderate" side and prevented the initial split from continuing. The same cannot be said of the Saharan affair, and in any case, the sides in the two conflicts of early 1976 largely coincided. The second incident was the alleged raid on the presidential palace at Cotonou, Benin, in which Morocco, Senegal and Gabon were supposedly implicated. The third incident was the invasion from Angola of the Shaba province of Zaire, in 1977 and 1978, where Moroccan troops have been sent to support the regime of Gen. Sese Seko Mobutu. The fourth was the establishment of the Rejection and then the Steadfastness Fronts in the Arab world, a split much more significant than the suspension of Egypt from the Arab League. All of these events dated from the latter half of the 1970s, but it took two more years before the summer of 1982 when they finally took organisational form with the OAU split.

There are two other influences on the leadership struggle besides the conscious - even if episodic - policies of the principle states. One is the role of the followers: As any active state knows, one's allies have a way of inserting their own interests and moving outside the assigned role of obedient supporters. If the leadership drive of a few states is a source of conflict in North Africa, the interests of their respective followers is a source of equal importance. The other is the role of external allies: As crises escalate, developing states find their own resources overtaxed, even when other developing states are enrolled in the conflict, and the natural tendency is to seek outside support from developed states. The highly principled campaigns to seek only "African solutions to African problems", expressed in other terms for other regions, derived its practical importance as a tactic to delegitimise the search for outside allies. Again, there is practical wisdom in the approach, since for developing countries, developed allies may provide power resources but they also bring "conflict by association" that is not always helpful to developing states' interests.

Position in Global Politics

The final source of conflict in North Africa builds on the previous factors and projects into the world beyond. (7)

The goal of the entire modern (post-medieval) history of North African societies (as elsewhere) in its broadest sense has been to achieve a better life for some or more of its inhabitants in contemporary terms. It was this drive which led rulers to look for new possessions and functions in the nineteenth century, in contract with the West (or to North Africa, the North). It was the broadening of this goal which the French called their civilising mission, and it is the same goal, with somewhat different specific components and a different cultural or ideological hue, which is associated with such ideas as development, modernisation, Islamic socialism, and others. Yet the concern is not merely one of domestic goals and policies, although it is primarily so. It also has foreign policy ramifications. On the field of foreign relations, it can also be said that a primary goal of developing states is to achieve acceptance in the world of the developed.

The philosophy and general concept of international relations have changed much since the years when they were a reflection of a Concert of Europe and a colonial system emanating from it. The major change can be summarised as the entry of new non-European states into that Concert (USA, Japan, possibly China), the self-destruction of the European centre and its replacement by a bi-polar security system, and the liberation of the formerly-colonial world from a subordinate to a legally equal status from which they may aspire to enter into the central Concert. This entry is above all a matter of generating and mobilising the elements of national power, and hence of development; in the past, the acquisition of these elements in sufficient quantity was demonstrated by successful participation in a major war, although there is no guarantee that that is the only way turning points in history may be marked in the future. But associated with this development, basically economic, is a choice of political strategies. Like readiness or worthiness for independence, readiness or worthiness for entry into the central Concert may not be immediately apparent to the naked eye and may have to be asserted, supported, and eventually imposed by political action.

There are three strategies for this action: New aspirants can simply compete and become accepted on their own, they can be coopted under an alliance or sponsorship of a member, or they can force their way in (on somewhat different terms) as the representative or leader of lesser outsiders. These strategies are as applicable to new firms entering into competition with established businesses, new graduates making their way into competition with lawyers or other professionals, or new blood making its way into a closed circle of established families, as they are to new-comers on the world arena, but are no less applicable to the latter context because of their universality. Historically, the states of North Africa, among others, tried the first strategy and failed, were pressed into the second in a parti-cularly rigid form, and now are faced with the three-fold choice again in different terms.

Tunisia is particularly apt as an example of a state which tried to hoist itself to the level of the developing European states in the middle of the nineteenth century, wavered in its strategy, and then fell into bankruptcy and receivership as a result. However, the same policy, with different dates and details, yielded the same result in Algeria fifty years earlier and in Morocco thirty years later. The borrowed status acquired under colonialism is less interesting since it does not concern national policy by North African polities; the policy of moving from proxy membership in the European concert to an independently-negotiated relation, however, is particularly well exem-plified in the shifting attitudes and negotiations with the European Community.

Morocco and Tunisia have concluded association agree-ments out of a provision for special consideration in the original Rome Treaty, whereas Algeria moved from de facto membership to its participation in the association agreement.

If at the present time, all three states appear to join in agreement on the strategy towards the EC, it is not as certain that the general strategy or all aspects of it toward entry into the developed centre will not be a source of conflict in the future. Again, the two major opponents are Morocco and Algeria, with Tunisia being closer to the Moroccan position and Libya to the left of Algeria. Morocco's approach - in a word - appears to be one of "joining 'em" while Algeria's is one of "licking 'em". Mo-rocco is more amenable to cooperating with developed - notably Western - powers, to benefiting from association with them, eventually to borrowing power from them to make up for its shortfall. Cooperation with France in Zaire, acceptance of US personnel on former American bases, a more flexible attitude in the Paris Conference on International

Economic Cooperation negotiations are recent examples. Moroc
co's strategy is a mixture of the first and second. Algeria
appears to be not only expanding its own capabilities both
by borrowing and reinvesting, but also has led the attack on
the Old International Economic Order while seeking to rally
smaller states behind it in the assault. Algeria's role in
calling the Sixth and Seventh Special Sessions of the UN
General Assembly and in leading the Group of 77 are important examples. Algeria's is a mixture of the first and third
strategies. Played to the fullest, these two divergent
choices of strategies are highly conflictual. That they have
not been so in the past is most likely evidence of the fact
that the battle has not yet been fully joined. The stage of
their own development does not yet qualify either state for
membership in the Concert of Developed Powers. As development procedes, however, the conflict - exacerbated by the
other causes described above - is likely to sharpen.

Libya's is a fourth strategy, more idiosyncratic and
less transferable but worthy of note if only because it is
so badly misunderstood. Qadhafi does not merely seek to
enter the Core Concert; he acts now as if he were already in
and uses his elements of national power to play as a
fullfledged actor. This refusal to "know his place" and the
frequent gap between role aspirations (or assumptions) and
means at hand make for an erratic performance that un-
sympathetic observers have mistaken for "craziness". Qadhafi's global policy is profoundly activist and revisionist.
He seeks to mobilise and revolutionise the Arab world for
the overthrow of imperialism, as he mobilised and revolutionised the Libyan armed forces for the overthrow of the
monarchy in the name of the people. Giving meaning to these
tactics is the ideology of the Third way, "the universalism
of Islam conceived as revolution". All means are mobilised
behind this task, but for Libya the most available and thus
the most evident means are financial and subversive, both
operating outside the "normal" rules of established diplomacy. Libya is therefore revisionist in means as well as
ends. Few follow its path, among the Third World or even the
Arab states, but many admire its gall.

THE SEA AS A SOURCE OF CONFLICT

Discussion can now turn to the Sea. The Mediterranean

has no new sources of conflict of its own, but is merely the extended arena - from the Maghrebi point of view - for land-origin conflicts to be pursued. (8) It can, of course, be an active or a passive arena, that is, it can bring its own elements to existing bases of conflict or simply be the place where conflicts take place. For example, there are boundary conflicts on the sea (or at least the territorial waters) as on the land, and there are different attitudes that govern Magrebi states' view of their role in Mediterranean security arrangements vis-à-vis the central (European) Concert. On the passive side, ideological differences may prevent Maghrebi states from cooperating on common interests on the Law of the Sea, for reasons having nothing to do with the sea itself. The following discussion will focus on the Mediterranean as an active arena, the passive aspect in a sense already being covered in the previous discussion. However, since many of the elements of an active arena are covered in other papers specifically addressed to them (e.g. energy, non-energy resources, boundaries, military activity), the discussion here will be indicative rather than specific.

It should be recognised that Maghrebi states stand in different relation to the Mediterranean. Algeria, Tunisi and Libya are Mediterranean states: All their ports are on the Sea, and all their commerce in the broadest sense must be Mediterranean before it becomes anything else (except for that which transits the far less hospitable Saharan routes to the south). Morocco is different: Its Mediterranan commerce is insignificant in comparison with its Atlantic activity, its Mediterranean relations are Atlantic before they are anything else (except for its Saharan commerce and its land access through Algeria, the latter frequently reduced or closed because of interstate conflict). Geographical imagery should not be overdone, but as long as perception is the screen for reality, it is of some relevance.

There are five areas of conflict related to the Sea as it concerns North Africa, some direct extensions of the previous six conceptual dimensions and others existing on more specific levels of their own.

Boundaries

The Sea knows its boundary conflicts like the land. There have only been two thus far, both involving the revisionist state par excellence of the region, Libya, but a few others are conceivable. One type of conflict involves the seaward extension of land boundaries between states, exacerbated on sea as on land by under-surface mineral deposits. The case at hand is the long maritime border conflict between Libya and Tunisia, bearing the characteristics of complexity: A land border not perpendicular to the shoreline, offshore islands, a concave coast, and offshore oil deposits. Two positive aspects of the conflict are noteworthy: First, it was resolved, after years of manoeuvering which could have escalated into a larger conflict involving outside powers, by appeal to the International Court of Justice and arbitration by international institutions. Second, the four conditions which made the dispute so intractable technically are present nowhere else along the Southern shore.

A second type of sea boundary dispute involves boundaries to national jurisdiction and territorial sea parallel to the coast. Here the case is again Libya, which has asserted its claim over the Gulf of Sirte as an inland sea, against the precepts of international law. This type of dispute is more properly a part of the discussion on the extension and delimitation of national water boundaries, but again it should be noted that, except for the even more complicated Gulf of Gabes, there is no other concave coast that could lend itself to the type of claim that Libya makes on Sirte. On the other hand, territorial waters can be the subject of troublesome claims, not only into the open Mediterranean, but more disruptively at the Mediterranean chokepoints bordered by Morocco and Tunisia. It would take a different mood on the part of both countries to raise a conflict with Spain and Italy, respectively, over the intervening channels.

A third form of sea boundary conflict involves the maritime implications of territorial claims, relating specifically only to one case, the anticolonial campaign against the Spanish islands and enclaves along the Moroccan Mediterranean coast. The Moroccan claims are simply on hold for the moment, until the Saharan question is resolved, but once Morocco's hands are freer, it no longer needs Spain's support or at least neutrality, and a national cause is required, the campaign against Ceuta, Melilla, and the islands will erupt. It is related in turn with the reasoning and outcome in the Gibraltar case, since the Spanish

168

enclaves are the mirror image of the British Rock. Use, possession, and extent of the territorial sea are matters that are involved in the recovery of the enclaves. A Falklands-like episode is not inconceivable at some distant time, despite the many obvious differences in geography and force levels, among others.

Bridges Across the Sea

Conflict may come over the bridges across the Sea. Such bridges are not numerous, and they have provoked no notable conflict as yet. A gasline has been completed from Hassi Rmel in Algeria through Tunisia to Minerbio in Italy; other pipelines are under consideration to Spain and other points. One might also consider the tanker routes with their expensive terminal facilities as commitments of similar magnitude as a permanent pipeline, although there is an important difference in degree. But the project of a bridge across the Straits of Gibraltar, which Morocco revived in 1982, is a more literal case of linkage that can also provide conflict.

The cooperative ingredient in such linkages is evident, and they are tangible cases of the functionalist notion that ties create interdependences that reduce conflict. At the same time, it should be noted that interaction and interdependence also create the ingredients of conflict. They raise expectations, bring otherwise conflicting parties into contact, restrain freedom of independent action, and provide the means of pressure and dispute which did not exist without them. The history of the gas negotiation between Algeria and France and Italy includes illustrations of all these effects, leaving bilateral relations in both cases closer but testier than before. However, as the cases also show, such conflicts are technical and diplomatic matters - elements in but not causes of a conflict.

The Sea as a Potential Battlefield

The Sea can also be a battlefield among Maghrebi states. The chances of a full-scale naval battle are slim in this day and age, to be sure, but curiously the Sea does have a role to play in the more typical warfare of our times - the guerrilla. In this, it has a history: The Mediterranean was a major (and at times the major) channel of support for the Algerian Army of National Liberation (ALN) during the Revolutionary War (1954-62), notably for supplies

169

and men from Morocco and from Egypt via Tunisia, and even a sophisticated French navy had difficulty controlling the flow. The war is over and the flow has stopped, but when conflict arises the same trails are used again. It was from the Sea that the airplane of unknown source dropped arms at Cape Sigli in 1978 for opponents of the Algerian regime. More important, arms from Libya to the Polisario have started to come by sea in 1981-82, transiting the Mediterranean and then the Atlantic to Nouadhibou and thence to the interior, as friction between Algeria and Libya reduces the flow across the "Polisario Trail" through the Sahara.

The Sea is useful for this kind of activity and not much can be done about it, even if it attained greater magnitudes. Hot pursuit on the high seas requires more of a navy than Morocco or Algeria has, although in the absence of effectiveness such naval action could produce an important political incident (as did the Cape Sigli airdrop). Beyond the incident, what is important in such a conflict is the dispute itself, and in this the Sea is only incidental. The day that Algeria decides to join loyally in the search for a political solution to the Western Saharan problem, Libyan diplomatic and military support and the sea channels for it will be a secondary consideration.

Bases and External Military Presence

The two remaining aspects relate to Maghrebi views and policies on external roles in the Mediterranean. Maghrebi states have little capability for projecting power into the Sea (as already noted), but they are able to offer other states an entry and an anchor for seapower on the shore. Through portcalls and base rights, littoral states grant access to outsiders and thereby establish an association which by its nature is conflictual with neighbours (unless they too have established the same association). Thus Tunisia's portcalls are more or less balanced between the US and the USSR, more frequent but for smaller ships for the Russians and slightly less frequent but for larger ships for the Americans; whereas Libya's and Algeria's portcalls are Russian. For these purposes, Morocco is not a Mediterranean country, since its major ports and portcalls are Atlantic.

Base arrangements are a less clear matter, because of the secret nature of the agreements; fortunately, consideration of the matter does not require up-to-date intelligence reports. The French, British and Americans left a

number of naval bases and port facilities for military use
along the Maghrebi coast after World War II and then the
decolonisation: Port Lyautey (Kenitra), Mers el-Kebir, Bi-
zerte, and Tripoli. Today, there is American and Soviet
access to the Moroccan and Libyan bases, respectively, but
no foreign presence or installations in the Algerian and
Tunisian bases. At the present time, these arrangements have
aroused comment and criticism from neighbours but little
else. They have not been used - nor are there plans to use
them - against neighbours or in connection with the conflicts that trouble North Africa. Nor does the neutral-
looking buffer of Algeria and Tunisia between Morocco and
Libya explain the lack of further naval-based conflict,
since the pattern is checkerboard rather than "Scandina-
vian", Algeria and Tunisia being more Soviet- and American-
leaning than their simple naval base arrangements would
indicate.

In a deeper sense, however, foreign bases and facilities are a sign of North African weakness, a foreign imposition on Maghrebi territory and an external power's penetration of the Southern shore. To make a fine but appropriate distinction, foreign bases give Maghrebi states protection but not capability, and they are of greater use to
the foreign power than to the host state. To be sure, the
host state expects to get equal value from the arrangement,
but the equal value is not in base-related military capability; it is rather in some form of "rent", such as the
economic aid which the US gave to Libya and gives to Morocco
or in favourable consideration of military supplies and
training which the US provides Morocco and the USSR provides
Libya. The base itself, however, is a platform for the
extension of foreign power, not an enhancement of host-state
power. In fact, even the element of protection may be thin;
foreign powers take up bases in foreign countries because
the site looks secure (and because it looks military useful,
of course), not because the foreign state wants to take on
its host's conflicts and cover them.

As a result, foreign bases may enhance conflict among
North African states but they are unlikely to serve as a
means of conflict. Indeed, foreign powers try to minimise
the concern that their bases cause for the host's neighbours, as the U.S. reassures Algeria that its Moroccan military facilities agreement for 1982 has nothing to do with
Algeria or the Saharan question, and the USSR assures

Tunisia that whatever arrangements it has with Libya are not directed against Tunis. But base relations make other problems harder to solve, as Algeria charges that Morocco is polarising the Saharan dispute, and they inhibit outside powers from being useful in conflict resolution since they give the host a means for restricting the outside power's latitude in mediation.

Perceptions of the Geostrategic Significance of the Mediterranean

Finally, in the broadest sense, the Sea is a source of conflict for Maghrebi states as they dispute - with words - its geostrategic place in the world. Whose lake is it? The views differ among the four states, and they merge into each state's own self-image and its concept of how to get on in the world. All share in viewing the Sea as properly belonging to its shoreholders, but the variations on the common theme are such that little basic unity remains.

The leitmotif of Moroccan global policy is equitable participation in regional affairs, with tolerance for diversity and protection against hegemony. Morocco therefore seeks entry into a concert of Mediterranean states, just as it participates in concerts of Arab and African states. In this regard, it maintains its special commercial agreement with the European Community and would like to participate in the Conference on Security and Cooperation in Europe as an interested Mediterranean state. Morocco's Mediterraneanness makes it part of the Concert of Europe.

Tunisia sees itself as a perpetual crossroads, centrally located in the Mediterranean. Its Mediterranean should be open, not a subject for access to the North but a terrain of commerce and intercourse. Such an open Sea need not exclude foreign fleets but it presumably would exclude foreign bases, at least on the Southern shore.

Algeria is the spearhead of Third World revisionism, more actively so under Boumedienne than under Benjedid but still holding the same philosophy. Its Mediterranean should be cleared of foreign (non-Mediterranean) fleets, although at one point it admitted that the Soviet fleet should be present as long as the American fleet was there (which it should not be). At the same time, eliminating superpower naval forces would also be a way of leaving the Sea with states more nearly equal in charge and thus giving a larger place to Algeria in regional politics.

For Libya the Mediterranean is part of the battlefield for the revolutionary forces of the Third World against imperialism, in which for the moment cooperation between the anti-imperialist Second and Third World leaders is required. Thus, in the name of the struggle against foreign forces, alliance with foreign forces is justified.

These four different attitudes factor into different stands on different aspects of the debate over the geopolitical place of the Mediterranean Sea, according to the way in which questions are posed. The result is different diplomatic positions and different types of demarches. But the outcome of this concern, as well as of the other four, is greater conflict with outside powers than among North African states. With the exception of the Libyan-Tunisian border dispute, for the moment apparently settled, none of the other specific instances or general categories of conflict has a serious Mediterranean aspect or extension, and none of the serious conflicts that separate Maghrebi states on the land - Sahara, Chad, Peace Process, NIEO, even the important general category of regional power rivalry - has an important ramification or theatre of operations on the Sea.

The reason is not hard to find: Not only is there little to fight over, but there is less to fight with. North African states are weak on sea, and weaker still in comparison with the naval power from the northern shore or from outside. Moreover (or therefore), North African states turn instead to the power vacuum which is unoccupied, the desert area which lies to the south and in which the power rivalries of North Africa are played out. To the extent that there is an association of outside powers and Maghrebi states across the Mediterranean, it is not so much to coordinate policy and increase power in the Sea arena as it is to provide mutual support in the Sahel/Sahara zone - the U.S. with Morocco in Western Sahara, France with Algeria in the region from Mali to Chad, the USSR with Libya in Chad and further southeast. This is all very tentative and perceptional, but it indicates a reverse direction of operations from conflict and cooperation in the Mediterranean.

THE SEA AS A DOOR THROUGH WHICH OUTSIDE CONFLICTS PENETRATE
THE REGION

The picture of conflict spillover into the Mediterranean from the Southern shore appears inconclusive. As a
result, it may be more insightful to turn the subject
proposition around and ask: What are the implications of
external activity for conflict escalation and deescalation
in North Africa? The two pairs of variables to be mani-
pulated are external power cooperation and conflict and
Maghrebi states' cooperation and conflict.

Relations with the Superpowers

As a general rule, North African states have always
hedged against polarisation of their own relations with the
superpowers. Morocco is armed by France and the U.S. and has
close political relations with the U.S., but has little
trade with the latter to compare with its 1978 $2 billion
phosphate agreement and $300 million fishing agreement with
the USSR. Algeria is armed by the USSR, with which it also
shares political views, but has little trade with the USSR,
its first trading partner being the U.S. and its second France. Tunisia, to be sure, is more exclusively Western-oriented, and Libya is completely Soviet-oriented on the political level, mainly Soviet-oriented in its military supplies,
but still Western-oriented in trade. Furthermore, the situation is fluid: Morocco has always had an eye toward maintaining Soviet relations, and Algeria in 1982 was moving to
diversify its military supplies with American and British
sources and reduce its dependency on the USSR. In Maghreb
relations, foreign ties act not so much as a limit as a
counterweight to other foreign associations - a "cante-
levered policy", as it were. This is a delicate situation,
and can be destabilising if one side withdraws support, as
the U.S. did in Libya. It limits outside powers' manoeuvrability and leverage for conflict resolution, and encourages escalation, although not necessarily on the Sea.

Breakdown of Regional Organizations

There is a total breakdown of organisational structure
and unity in the international relations of the region, but
existing organisations played such a limited conflict resolution function that the impact on regional conflicts is
slight. The collapse of the OAU in July-August 1982 into two

camps mirrors the division of the Arab League in 1977 (both around the twentieth anniversary of the respective (body); both have weakened organisations whose capacity for conflict management was very low already; they will be missed for other reasons, until reconstituted, but scarcely for the impact they have had on Mediterranean politics.

Superpower Hegemony and Conflict Resolution

A predominant greatpower presence in the Mediterranean facilitates conflict resolution. It reduces the chances of littoral states' playing off one outside states against another and it reduces distracting competition from another Cold War state. The U.S. could better mediate the Camp David agreements after the Soviet Union was removed from the immediate area; conversely, the chances of reducing conflict in the Horn of Africa in the late 1970s were eliminated by the primary need of the great powers to keep their own balance while shifting sides during the Ogaden war. It would have been more difficult for the U.S. to rein in Libya in various theatres during 1981 if the Soviets had had a greater naval presence in the Gulf of Sirte. Competition among various Western powers is less important, and can even provide alternative leverage if the states will concert on specific policies from time to time.

The potential for Soviet Penetration

Predominant Soviet penetration of a North African state (Friendship Treaty, Karmal coup, for example) would be highly disruptive to Western predominance in the Mediterranean and would make external efforts at conflict management difficult, but would also isolate the penetrated state. No such possibility appears on the horizon; neither a toppled monarchy in Morocco nor a stumbling succession in Tunisia nor a military radicalisation in Algeria - the likely "worst futures" for those three states - suggest a dominant Soviet penetration, just as in Libya the forces threatening the Qadhafi regime come from the dissident bourgeoisie rather than from the radical left. A Soviet-Steadfastness Front alliance would be the most likely entry for penetration, but the Soviet Union has not appeared very steadfast as a supporter of Palestinian revolution, where there is some mileage to be gotten. It is even less likely to expose itself by espousing some less promising cause in an intra-Maghrebi conflict.

Two Models for the Future

Putting all these elements together in a concluding attempt to project into the future, two model scenarios stand out strongly, both in the abstract and in analysts' discussions. One is a cooperative future, essentially a model for North African unity. According to this model, the causes of conflict dissolve as the initial needs of state-and- nation-building are met, and Maghrebi states surmount their inward national focus through regional cooperation, both economic and eventually political. As they turn to each other, they turn against the outside world and refuse the offers of interference extended to them by West and East alike. The other is a conflictual future, an overriding checkerboard pattern. Here the causes of conflict would take on a self-sufficiency of their own, destroying any of the attenuating tendencies inherent in Maghrebi society, and would become the dominant mode of North African relations. In the process, outside interference would become more pronounced and the checkerboard would be internationalised, making North Africa – and with it, the Mediterranean – an arena of interlocking global conflict, a new Horn of Africa.

Neither of these scenarios seems likely. Each neglects the logic and reality of the other, and carries with it consequences that are therefore unreal. Neither Maghrebi unity nor Cold War implantation fit the characteristics previously described. More likely is a third and less integral scenario of mixed characteristics, a future of attenuated conflict and interrupted cooperation, oscillating between the two poles but never attaining either because of constraints and remanents of the other. Although this represents a continuation of the past patterns of relations, its specific forms obviously are shaped by the particular events and contexts of the future. Its Mediterranean impact is likely to be minor, however, involving neither a major attempt to evacuate the Sea of its foreign navies, nor an offer of bases to Cold War powers, nor a serious spillover of Maghrebi conflict into the water.

NOTES

1. Much of this chapter is based on a broad range of general works on North Africa rather than being specifically attributable. See I. William Zartmen (ed.), Man, State and Society in the Contemporary Maghreb, Praegar, New York, 1973 and bibliography; Clement Henry Moore, Politics in North Africa, Little Brown, Boston, 1970; Abdullah Laroui, L'Histoire du Maghreb, Maspero, Paris, 1970, (English translation: Princeton University Press, Princeton, 1977); I. William Zartman, Ripe for Resolution: Conflict and Intervention in Africa, Yale University Press, New Haven, 1983.

2. On Maghrebi boundaries, see John Damis, Conflict in Northwest Africa, Hoover, Stanford, 1983; Francisco Villar, El Proceso de Autodeterminacion del Sahara, Fernando Torres, Valencia, 1982; I. William Zartman, International Relations in the New Africa, Prentice Hall, Englewood Cliffs, 1966; Robert Rezette, Le Sahara occidental et les frontieres marocaines, Nouvelles Editions Latines, Paris, 1977; Maurice Flory, "La notion du territoire arabe et son application au problème du Sahara", Annuaire Française du droit international, No. 3, 1957, pp. 73-91.

3. In addition to works listed under footnote 1, see Elbaki Hermassi, Leadership and National Development in North Africa, University of California Press, Berkeley, 1972, and specific works such as John Waterbury, The Commander of the Faithful, Columbia, New York, 1970; Remy Leveau, Le fallah marocain, defenseur du trone, FNSP, Paris, 1976; William B. Quandt, Revolution and Political Leadership: Algeria, MIT Press, Cambridge, 1969; Charles Micaud et al, Tunisia: The Politics of Development, Praeger, New York, 1964; Ruth First, Libya, Penguin, Baltimore, 1972; Maurice Flory (ed.), La Libya nouvelle, CNRS, Paris, 1975.

4. See I. William Zartman (ed.), Political Elites in Arab North Africa, Longman, New York, 1982; Michel Camau, La nation de democratie dans la pensée des dirigeants maghrebins, CNRS, Paris, 1971.

5. See Zartman, International Relations ... and "North African Foreign Policy", in L. Carl Brown (ed.), State an Society in Independent North Africa, Middle East Institute, Washington, 1966.

6. See Rene Otayek, "La Libye revolutionnaire au sud du Sahara", Maghreb-Machrek, No. 94: (October 1981), pp. 5-35.

7. In all the North-South and foreign policy literature, there is little that analyses and compares the various
global role images of developing countries, important as
this theme is. See K. J. Holsti, "National Role Conceptions
in the Study of Foreign Policy", International Studies
Quarterly, Vol. XIV, No. 3, (September 1970), pp. 233-309.

8. See I. William Zartman, "The Mediterranean: Bridge
or Barrier?", US Naval Institute Proceedings, Vol. XCIII,
No. 2, (February 1967), pp. 63-71.

Chapter Six

SUBNATIONAL CONFLICT IN THE MEDITERRANEAN REGION

Brian M. Jenkins

Almost since the beginning of Western civilization, the Mediterranean Sea has been a theatre of conflict, a boulevard for battleships and invading armies, a moat between great empires, diverse cultures, religions, and political systems. What the nations of the Mediterranean have in common, besides a shoreline, is 25 centuries of warfare – warfare between the Romans and the Carthaginians, between Moslems and Christians, between the corsairs of the Barbary Coast and the merchant fleets of the European kingdoms, between the weaker non-European countries and the colonial powers, between revolutionary regimes in North Africa and the Middle East and the industrial nations of Western Europe.

PATTERNS OF CONFLICT IN THE MEDITERRANEAN

Since the beginning of the twentieth century, there have been 49 wars, revolutions, and other armed conflicts involving the countries of the Mediterranean, 36 of them since World War II. (See the chronology at the end of this chapter.) In the 16 conflicts which have occurred during the last ten years, 10 of the 18 nations that border the Mediterranean were directly engaged, and three more were indirectly involved.

In reviewing this chronology, one can see several persistent conflicts that repeatedly erupt into open warfare. For most of the twentieth century, Jews and Arabs have fought for control of historic Palestine. With the creation of the State of Israel and the development of a Palestinian nationalist movement, the struggle changed form but continued. Fourteen of the 47 entries in the chronology are

179

related directly or indirectly to this struggle. Territorial disputes in the Western Sahara have been the source of fighting between Spain and Morocco and Morocco and Algeria since the 1950s. Communal disputes between Greeks and Turks on the island of Cyprus and between the Greek and Turkish governments over control of Cyprus have caused violent conflict since the 1950s. Friction between the Christian and Moslem communities in Lebanon has erupted into civil war twice since the 1950s and is the source of continuing armed conflict in that country. Radical Arab regimes in Algeria, Libya, and Syria have fought with or attempted to subvert more conservative Arab regimes in Morocco, Tunisia, and Egypt, and have provided support to terrorist groups in Spain, France, and Italy. Moslem fundamentalists have challenged secular regimes in Egypt and Syria. Separatist movements have occasionally appeared in Sardinia and Corsica. These patterns of conflict seem likely to continue.

Many of the conflicts in the region are internal (for example, the 1980 revolt in Syria), and many of the conflicts between nations also involve subnational or nongovernmental forces as we see in the complicated situation in Lebanon. (This does not count the terrorist campaigns in several Western European nations.) This fragmentation of warfare towards more numerous limited conflicts involving both national and subnational forces is part of a worldwide trend.

In many respects, the future face of war is revealed in the course of armed conflict in Lebanon since the late 1960s. Warfare in that country has continued on three levels: conventional war, guerrilla warfare, and terrorism. The conflict is concurrently a war among nations, a war between the State of Israel and a powerful nonstate actor – the Palestine Liberation Organization, a war among Lebanese factions, and a multitude of terrorist campaigns. It involves regular armies, guerrillas, private militias, terrorist gunmen and bombers, some of whom are openly assisted or covertly sponsored by foreign governments, by political or religious factions, and even by other terrorist groups. The conflict in Lebanon is likely to be representative of armed conflict in the last quarter of the twentieth century: a mixture of conventional warfare, guerrilla warfare, and campaigns of terrorism, openly fought or secretly waged, often without regard to national frontiers, by armies as well as irregular forces, directly or indirectly.

THE CRADLE OF INTERNATIONAL TERRORISM

The disputes and conflicts among and within the nations around the Mediterranean Sea have manifested themselves in a high level of terrorist activity. Thirty-six percent of all recorded incidents of international terrorism between 1968 and the end of 1981 occurred in the 18 Mediterranean countries. The same pattern continued in 1982 with 37 percent of the total number of incidents occurring in the region. No other geographic region in the world has experienced such a high level of terrorist activity. This terrorist activity is not distributed evenly among the 18 nations. Some of the countries, like Albania, Algeria, Libya, Malta, Morocco, and Tunisia, have suffered very little or no international terrorist activity within their borders. (Yugoslavia, while relatively free of terrorist activity at home, has been a frequent target of Croatian terrorist activity abroad.) Seven countries - France, Greece, Italy, Israel, Lebanon, Spain, and Turkey - account for 94 percent of the activity within the region and 34 percent of the total volume of international terrorist activity worldwide. (1)

The Mediterranean is the cradle of international terrorism in its contemporary form. Terrorist tactics have, of course, been used for centuries by both governments and revolutionaries. It is, however, in the postwar struggle in Palestine and in the guerrilla campaigns against the colonial powers that we first find campaigns of deliberate terrorism. Terrorist tactics were used extensively by the Irgun and later the Stern Gang as well as by their Arab adversaries in Palestine during the 1930s and 1940s. (2) They were used by Greek nationalists of the EOKA-B fighting the British in Cyprus during the 1950s. (3) They were used by the FLN in its struggle against the French, as well as by the OAS as part of its campaign to keep Algeria French. (4) It is out of these struggles that the ideological and doctrinal foundations of contemporary terrorism emerged.

Colonial insurgents defined colonialism itself as "violence in its natural state, and thus the only possible means of ending it was by greater violence". (5) Greater violence was not only rationalised by the colonial insurgents, but the legitimate targets of violence were potentially broadened to include the entire colonial machinery: government officials whether highranking dignitaries or

Colonial insurgents defined colonialism itself as "violence in its natural state, and thus the only possible means of ending it was by greater violence". (5) Greater violence was not only rationalised by the colonial insurgents, but the legitimate targets of violence were potentially broadened to include the entire colonial machinery: government officials whether highranking dignitaries or minor bureaucrats, whether civilian or military, policemen, plantation owners, colons, indigenous collaborators, just about anybody who participated in the colonial structure; which in its extreme could mean anybody who did not actively participate in the struggle to overthrow the colonial rulers.

In their pursuit of independence, the Algerian nationalists carried their struggle to the French metropole to maintain an impression of war in France as well as Algeria. While terror was employed by all participants in the Algerian conflict – by the FLN against Algerian and French targets in Algeria and France, by the French Army particularly in the battle for Algiers, by the OAS against French and Algerian targets in Algeria and France, by the barbouzes (secret French agents) against the OAS – it is the FLN that provided inspiration and instruction to the Palestinians in their struggle against Israel. The victory of the FLN represented a triumph of Arabs over a Western power. A number of Palestinian leaders went to Algeria in the early 1960s where they met FLN leaders whose doctrines and tactics were later to influence the development of the Palestinian movement, in particular, Fatah. (6)

Frustrated by the failure of Arab military power in the Six Day War of 1967 and by a world that ignored their plight, a new generation of Palestinian groups launched an international campaign of terrorism designed to bring them worldwide attention. (7) The Palestinians did not confine their operations to Israel and the Occupied Territories but carried their terrorist campaign abroad, primarily to Western Europe, where they struck Israeli, Jewish, European as well as Arab targets. This was consistent with the Palestinians' view that Israel was a "colony" maintained by Zionists and imperialists just as Algeria was a colony of the French. In an effort to increase their capabilities and broaden their struggle, the Palestinian groups developed contacts with other subnational groups in Europe and the Middle East, such as Germany's Red Army Faction, ETA in Spain, the Turkish People's Liberation Army, and later the

Red Brigades in Italy and Direct Action in France, and provided them with various forms of support.

Out of the Palestinian training camps and the civil war in Lebanon new groups emerged like the Armenian Secret Army for the Liberation of Armenia (ASALA), closely associated with the Popular Front for the Liberation of Palestine, and Black June or Al-Assifa, a splinter of Fatah. Both groups have operated worldwide.

Nations of the region thus face terrorism on two fronts: terrorism by indigenous groups using terrorist tactics to obtain independence, overthrow the government, or force it to adopt authoritarian measures, and, foreign or ethnic-based groups carrying out terrorist operations on behalf of foreign causes against foreign as well as indigenous targets. The two often overlap. In the former case, the problem is primarily local. In the latter case, terrorist activity cannot be isolated from other modes of armed conflict among nations or within other nations, but can be seen as another dimension of warfare. Thus, it is not simply that the countries of the Mediterranean individually suffer high levels of terrorist activity. Much of the terrorist activity of the region is inextricably intertwined with international conflicts.

With technological advance and economic progress, has come an increase in movement across borders of persons, news, ideas, money, goods - and conflict. The people come as tourists, students, technicians, workers, exiles, and refugees. Modern communications, in particular the mass media, keeps them informed about the quarrels of their homeland. A quarrel in one part of the world increasingly produces violent echoes elsewhere. Conflict in one country often also has effects in other countries. It may imperil investments, cut off sources of vital material, eliminate markets, and put the lives of diplomats or citizens living in another country at risk. One aspect of interdependence is the increased internationalisation of conflict.

A LABYRINTH OF SECRET WARS AND SECRET DEALS

Governments have increasingly supported or exploited guerrilla and terrorist groups as a means of waging surrogate warfare against their foes. It may be considered impolite to mention these episodes, but all incidents cited here have been widely reported, are hardly denied, and illustrate a worldwide trend toward exploitation of terrorism by governments. Algeria has provided support in the form of training for Spain's Basque terrorists and France's Corsican separatists, and also supports Polisario guerrillas in their fight with Morocco. Spanish authorities note, for example, that in 1976, approximately 100 members of ETA (politico-militar) received basic military training at the Police Academy in Algiers. (Basque separatists also have received training in Palestinian camps in Lebanon.) (8) According to another source, Corsican and Breton separatists were trained at Blida military camp southwest of Algiers. (9)

Libya has provided money, training facilities, and asylum to Basque, French, Italian, and Palestinian terrorists. Libya has provided support to Polisario guerrillas. Libya has also backed a plot against King Hassan of Morocco (who provided asylum for opponents of Colonel Qadhafi) in 1971, supported a plot against President Sadat of Egypt in 1974, and sponsored a guerrilla raid into Tunisia in 1980. (10)

Syria supports a number of Palestinian groups including Al Assifa, a splinter group of Fatah which has carried out terrorist operations in Lebanon and Western Europe. Spanish, Italian, French, Turkish, and Armenian terrorists have received training at PLO camps in Syria and Lebanon.

Israel has provided support in the form of weapons, money, and training to Phalangist militias in Lebanon. Israel also may have approached the Red Brigades with an offer of assistance. It is not clear what the Israelis might have been after in the latter case. Perhaps they simply were attempting to garner information about the Palestinians who, it is known, had given the Red Brigades and other terrorist groups in Europe weapons and explosives with an understanding that a portion of these would be set aside for future Palestinian operations.

While not actively cooperating with terrorists, governments have arranged quiet deals with terrorists in return for immunity from further attack. Held hostage by the Red Brigades, Aldo Moro revealed that Palestinian terrorists apprehended in Rome were released as part of a deal with the Black September Organization. (11) According to subsequent revelations in the press, this kind of thing has occurred on several occasions. There are reports that Italy at present has similar tacit arrangements with Fatah, the PFLP, and

While not actively cooperating with terrorists, governments have arranged quiet deals with terrorists in return
for immunity from further attack. Held hostage by the Red
Brigades, Aldo Moro revealed that Palestinian terrorists
apprehended in Rome were released as part of a deal with the
Black September Organization. (11) According to subsequent
revelations in the press, this kind of thing has occurred on
several occasions. There are reports that Italy at present
has similar tacit arrangements with Fatah, the PFLP, and
Al-Assifa. It has been alleged that several other Western
European governments had tacit deals of this sort - prompt
release of Palestinian terrorists apprehended in return for
immunity from further terrorist attacks. Both Spanish and
Italian officials have complained at times about the unwillingness of French authorities to move vigorously against
Basque and Italian terrorists resting up in France. (12)

French officials allegedly were also involved in arms
traffic with the Basque terrorists. According to published
accounts, French intelligence officials did business with
ETA in order to smuggle weapons through Spain to French-
speaking countries of northern Africa. (13)

A number of states have also directly adopted
terrorist tactics themselves, sending teams of assassins to
silence foreign foes or domestic opponents living abroad. In
response to terrorist attacks on Israeli targets abroad,
Israeli agents assassinated a number of Palestinians in
Europe believed involved in terrorist activities. Syria is
believed to have commissioned the gunmen who assassinated
that country's former premier in Paris. Libya openly avowed
its campaign directed against Libyan "traitors living
abroad" and was accused of sending teams to kill American
diplomats in Europe. The Spanish have been accused of
operating a "parallel police force" in France dedicated to
killing leaders of the Basque separatist movement. Yugoslavia has been accused of killing emigre Croatian exiles.
Outraged by continuing Armenian terrorist attacks against
Turkish diplomats, Turkish officials have recently warned
that there would be no sanctuary for the Armenian gunmen,
implying direct extraterritorial action.

Governments have dealt with terrorists in the last
quarter of the twentieth century much as the European powers
dealt with the pirates that plagued the Mediterranean in the
early seventeenth century. They have simultaneously
tolerated, combatted, fomented, supplied, and exploited them.

The European navies could have crushed the pirates if they had been willing to cooperate and cared to make the investment, but the pirates only molested trade, they did not interrupt it. Navies were saved for larger contests. Moreover, the pirates were trading partners. At least until the early eighteenth century, the corsairs of the Barbary Coast bought most of their guns and a large portion of their gunpowder from European merchants, and Europeans bought a number of commodities from the North African states - a profitable trade for some. European renegades trained private gunners. And at times, European kings sought alliance with the corsairs to prey upon the shipping or distract the navy of a rival European power.

So it would be equally incorrect today to view the posture of governments, specifically the governments of the Mediterranean countries, or even the governments of those countries that have suffered the most from terrorism as one of unmitigated hostility towards their adversaries. Instead, just beneath the rhetoric, one enters a labyrinth of secret wars and secret deals, of direct action and deliberate inaction.

TERRORISM IN THE MARITIME ENVIRONMENT

In recent years, a number of security analysts have warned that terrorists who previously had operated only on land or, considering hijackings, in the air, may turn to targets in the maritime environment. Their reasoning goes that with the increasing importance of offshore facilities to the world's supply of energy, the vulnerability of energy-related marine assets, and inspiration provided by fictional offerings such as The French Atlantic Affair, Shipkiller, The Devil's Alternative, Seawith, and Firestorm, terrorists will inevitably turn their attention to harbours, offshore platforms, and ships at sea. (14) In support of this thesis, it could also be said that because of the requirements to capture headlines, terrorists are under constant pressure to do something new, and that while terrorist tactics have changed little, the spectrum of targets attacked by terrorists has steadily expanded. Attacks on targets in the marine environment would provide needed novelty. Moreover, security specialists have in the last five years noted an increase in criminal activity directed against ocean industries.

Certainly, there are ample maritime targets in the Mediterranean. Thousands of vessels - oil tankers, LNG carriers, cargo ships, passenger liners, fishing boats, ferries, naval craft, and pleasure boats - sail its waters daily making it one of the world's most heavily trafficked bodies of water. Its numerous harbours contain oil and LNG terminals, refineries, regasification plants, and huge petrochemical installations. Underwater pipelines and cables cross its bed. In recent years, offshore installations for the exploration and production of hydrocarbons and seabed minerals have proliferated and are of growing economic importance, as is pointed out in Chapters One and Two of this book.

ARE MARITIME TARGETS ATTRACTIVE TO TERRORISTS?

Although maritime targets may be vulnerable, are they in fact attractive to terrorists? Certainly the seizure of a passenger liner, loader tanker, or offshore platform would be novel enough to capture world headlines, and could give the terrorists some leverage. The destruction of a refinery or terminal at a major port, the sinking of a ship blocking a harbour entrance, or the sinking of a naval vessel have high political value for the terrorists and could impose severe economic losses on the opponents. Extortion involving threats against ships at sea, offshore platforms, or critical facilities - a growing area of ordinary criminal activity - could replace income derived from bank robberies, ransom kidnappings, and foreign patrons. Terrorist groups need large sums of money. Bank robberies entail great risks. Kidnappings are becoming more difficult and ransom payments in general are declining. Corporations faced with credible threats might be inclined to pay off rather than risk critical and costly facilities. Terrorists may also see in maritime targets a possibility to wage economic warfare by attacking port facilities, offshore platforms, shipping, or carrying out actions that may cause great ecological damage. Finally, as they have done in the nuclear domain, terrorists could attempt to exploit environmentalist sentiments, attacking, for example, ships carrying nuclear waste or contaminated soil.

There are also certain constraints. Port facilities,

187

offshore platforms, and ships at sea may be theoretically vulnerable but the ease with which moving ships can be boarded, platforms taken over, or refineries set on fire should not be exaggerated. Terrorists are not, for the most part, highly trained commandos. Taking over ships or platforms may require more men than most terrorist groups generally have fielded. Authorities can more easily cut off communications and isolate a ship or offshore platform from television cameras and all the paraphernalia of modern mass media than they can an embassy in a capital city. Holding a large ship or oil platform is more difficult than holding a building. In short, increased terrorist attacks on maritime targets are not inevitable.

Although subnational groups have not operated extensively in the maritime environment, they have carried out a variety of actions: attacks on port facilities, sabotage or seizure of freighters and tankers, seizures or sinkings of ocean liners. Some examples worldwide:

January 1961: Seventy men armed with machine guns and hand grenades seized control of a Portuguese liner with 600 passengers on board. Opponents of the Portuguese government, they demanded political recognition of "this" liberated part of the national territory. Negotiations ended the episode 11 days later.

June 1971: PFLP terrorists carried out an assault on the Liberian-registered oil tanker "Coral Sea". Terrorists on a speedboat fired 10 bazooka shells at the tanker, causing some damage but no casualties. The attack occurred in the Strait of Bab el Mandeb at the entrance to the Red Sea. It was intended to deter tankers from using the Israeli port of Eilat on the Red Sea.

August 1972: Black September Organization set fire to an oil storage facility at the port of Trieste because it supplied oil to West Germany and Austria, both of which, the BSO said, supported Israel. The fire caused $7 million in damage.

December 1972: A plan by Black September Organization terrorists to hijack an Italian passenger ship between Cyprus and Israel was frustrated by police.

March 1973: A Greek charter ship carrying 250 tourists bound for Haifa sank in Beirut harbour following an explosion. There were no casualties. Black September claimed credit for the incident.

February 1974: Three gunmen seized control of a Greek freighter in the port of Karachi. They threatened to blow up the ship and kill their hostages unless the Greek government freed two imprisoned Arab terrorists. Greece agreed to commute the sentences of the two and the ship hijackers were flown to Libya.

May 1975: Oil tankers in the Persian Gulf were alerted to reports of a plot by Arab skindivers to hijack a ship, the General Council of British Shipping said.

August 1976: A Greek vessel in Lebanon was sunk by three limpet mines. The attackers were believed to be members of a right wing Lebanese Christian group. The vessel was partly loaded with cargo believed to be arms destined for the Al Fatah organization.

November 1977: The Algerian-backed Polisario Front guerrillas attacked a Spanish trawler with mortar and machine-gun fire, and seized three Spanish fishermen. The guerrillas used a rocket- armed inflatable speedboat in the attack. The guerrillas claimed they had seized the trawler because it had "violated the waters of the Sahara republic to pillage its maritime riches". The guerrillas announced on November 25 that they would release the three Spanish fishermen. On December 23, eight French hostages who had been held captive by the Polisario Front were handed over to UN Secretary General Kurt Waldheim by a minister of the Polisario Front at the UN office in Algiers.

September 1978: The sinking by the Israeli Navy of an explosive-laden freighter foiled a sensational Al Fatah terrorist plot. The terrorists planned to sail the vessel into the Sinai port of Eilat, firing 42 122-mm rockets at the port's tank farm and then ramming the 600-ton boat, crammed with more than four tons of explosives, onto the crowded beach.

July 1980: Police arrested three Corsican separatists of the Front for the National Liberation of Corsica (FNLC) who were planning to blow up oil tankers at Fos, the Marseilles Oil Terminal, and nearby oil refineries. Police officials said that 44 pounds of explosives had been found attached to a pipeline at the Berre refinery near Marseilles.

October 1980: A Libyan ship in for repairs at the port of Genoa almost sank following the explosion of a device that had been attached to the hull below the waterline. The Maltese National Front was suspected of the attack.

These events provide an idea of the range of targets: passenger liners, freighters, tankers, a trawler, port facilities (primarily those associated with the transfer or refining of oil). Except for their size and inherent strength, these things are virtually unprotected. The events that have occurred also provide an idea of the adversaries' techniques and weapons: limpet mines attached by scuba divers, rockets fired from small speed boats, an explosives-filled freighter armed with 122 mm rockets.

The PFLP's attack in the Coral Sea in June 1971, along with the August 1972 fire at the oil storage facility in the port of Trieste, were part of a Palestinian campaign aimed at oil supply lines in general and, specifically, oil being shipped through the Israeli pipeline from Eilat to Ashkelon. The Palestinians were particularly incensed that oil from the Persian Gulf was being transported by sea to Eilat, off-loaded and pumped 153 miles through the pipeline to Ashkelon where it was then shipped by sea again to the Black Sea ports and Trieste, thus avoiding the longer route around the African Cape. This arrangement, the Palestinians said, violated the Arab boycott of the Jewish state. Between 1971 and 1973, Palestinian commandos also carried out numerous attacks on pipelines and storage facilities in Jordan, Lebanon, and Saudi Arabia, as well as attacks on oil and gas facilities in West Germany and the Netherlands. The Arab oil embargo negated the need for further attacks.

We have seen few examples of sustained guerrilla war at sea. The American-backed secret war against Cuba in the 1960s comes to mind. In that case, anti-Castro Cuban exiles were trained in navigation techniques, and were provided with small, fast boats, recoilless rifles, machine guns, and mines. For several years they carried on a naval guerrilla war, sinking ships off the coast of Cuba, raiding Cuban ports. The military effect of this campaign was probably negligible, but a similar campaign in the Mediterranean Sea where commercial traffic is heavier, the risks of ecological damage greater, and the political situation more complicated, could cause some problems.

CURRENT AND FUTURE CAPABILITIES

Subnational groups active in Mediterranean countries currently possess a limited capacity for low-level maritime operations. Weapons move by sea from Western Europe to North Africa back to Western Europe. Weapons also move from Bulgarian ports on the Black Sea to ports in Cyprus, Lebanon and Syria. The clandestine trade in guns is only part of a larger contraband trade that includes narcotics and other controlled goods moving back and forth across the Mediterranean. This trade is carried out by small freighters and a vast flotilla of smaller, faster boats able to avoid the detection of customs authorities.

Of all the subnational groups active in the region, the Palestinians possess the most developed maritime capability. Fatah has a small "navy" of markabs (500-ton vessels) operating out of Cyprus and the Lebanese port of Tripoli, and sailing under various flags. The Palestinians have used these as "mother ships" from which they have launched fast boats and dinghies to land commandos and weapons on the Israeli coast. The 1975 seizure of hostages at the Savoy Hotel in Tel Aviv and the 1978 attack on a tourist bus north of Tel Aviv were carried out by Palestinian guerrillas who came ashore this way. In 1982, there were unconfirmed reports that Fatah had received four one-man submarines from Yugoslavia similar to the one-man submarines employed in World War II. Both the PFLP-GC and the PNF have Frogmen, reportedly trained in Yugoslavia. They are probably capable of operating underwater with open circuit systems (breathing air compressed in tanks, expelling the exhaust into the water). More advanced closed circuit systems recycle the exhaust and are preferable for military purposes not because they improve the divers' range or length of time underwater, but rather because they do not leave a revealing trail of bubbles on the surface. Closed circuit systems, however, cannot be used below 30 feet under the surface owing to the danger of oxygen poisoning.

In terms of explosives, limpet mines have been used by Palestinians as well as found attached to the hulls of ferries running between Spain and Morocco. An interesting use of technology is provided by the Basques who used an explosives-filled, radio-controlled model boat to damage a Spanish naval vessel.

Looking to the future, the growth of an offshore industry will bring with it a corresponding growth in offshore support and service industries and an increase in the number of helicopters, trained commercial divers, and manned and remotely-operated submersible vehicles. As of mid-1982, there were 58 manned submersibles home-ported in Mediterranean countries. (16) Future adversaries, be they terrorists or ordinary criminals, may be able to acquire or draw on these specialised skills and technology to carry out operations against maritime targets.

PROSPECTS FOR COOPERATION

While subnational conflict in the region is not a maritime problem but a political one, many of the events described, had they occurred in the Mediterranean Sea, would have had major international consequences. Certainly the hijacking or sinking of an ocean liner filled with passengers, or the hijacking of a loaded oil tanker at sea or in harbour could pose a complex problem involving the governments and possibly the naval forces of several Mediterranan nations.

Looking to the future, naval forces may have to review their capabilities for protecting vital port facilities and shipping from seaborne terrorists, and rescuing hostages held aboard hijacked vessels. Governments may have to examine how they would cooperate in a major incident at sea involving the interests of several nations.

What are the prospects for cooperation? If terrorists were something in the water, the Mediterranean nations might cooperate to combat them as they have agreed to act collectively in cleaning up pollution. But as terrorism is an effluent of the region's many and diverse conflicts, a collective response in this domain seems unlikely. There is at present no Mediterrranean political forum. This has not prevented some cooperation among some governments in some areas, but overall that cooperation has been quite limited.

According to information provided by the United States Department of State, all but three of the 17 Mediterranean nations (Albania, Algeria, and Malta) have signed and ratified the three international conventions on airline hijacking (the Tokyo Convention of 1963, the Hague Convention of 1970, and the Montreal Convention of 1971). These conventions certainly have not prevented airline hijackings, but they facilitate cooperation when one does occur. Generally, it has been easier to obtain international agreement on specific terrorist tactics than it has been to obtain agreement on dealing with terrorists.

The Council of Europe, of which seven of the Mediterranean countries are members and two more have observer status, has tackled the problems of terrorism on several occasions. The first effort led to a European Convention on the Suppression of Terrorism (1976). The Convention takes the pragmatic approach by enumerating specific offenses for which the normal political exception to extradiction will not apply. The results, however, were disappointing. Although all of the 21 member states, except Ireland and Malta, signed it, as of June 1982 less than half of them had ratified it.

In 1980, the Council convened a conference at Strasbourg which took a much broader approach to the problem of terrorism. The Strasbourg conference and subsequent meetings produced no new convention but did contribute a number of specific suggestions aimed at combatting terrorism with such measures as public education and the exchange of information among member countries. (17)

In order to maintain the broadest possible consensus, the Council of Europe focused its efforts on how to combat its domestic terrorist groups. It set aside the vexing issue of external support for terrorism. This avoided problems arising from European dependence on oil from the Middle East and other economic interests in Mediterranean countries, European sympathies for the plight of Palestinians, and its desire to avoid blurring the problem of terrorism in democratic societies with the use of terrorist tactics by guerrilla groups in the Third World. Even then, it has been difficult to gain agreement.

It is hard to imagine a cooperative effort among the Mediterranean nations except that which might be hastily nailed together in a crisis situation that involves the interests of several nations. The cooperative effort involved in the withdrawal and relocation of the PLO is an example. It certainly did not solve the Palestinian problem, but it probably did save a great many lives.

Individual national efforts to provide security for maritime targets in the Mediterranean and reaction to any episodes that may occur do not seem to be as well developed as they are in the North Sea countries (Netherlands, Norway, United Kingdom) who have developed contingency plans and specialized units for dealing with the threat.

THE OUTLOOK

For the foreseeable future, the security of ships at sea and offshore facilities will remain primarily the responsibility of ship owners and platform operators with incidents handled on an ad hoc basis.

As for the future, continued international and subnational conflict seems likely to occur in the region. It remains to be seen whether any new government in Lebanon will be able to mend relations between the Christian and Moslem communities, and continued fighting between Christians, Moslems, and Druzes seems likely. Israel and the Arab countries remain at odds. The Palestinian problem has not been solved.

Their defeat at the hands of the Israelis and dispersal to various countries around the Mediterranean give cause for renewed terrorist activity and make central control and constraint by more moderate elements more difficult. Their presence in North African countries may also facilitate infiltration into Europe. A new coalition of Palestinians, Lebanese, Leftists, and Armenians may be forming with its headquarters on Cyprus. Early reports indicated that Greece would turn a blind eye to its activities. Cypriot officials reportedly deny that Cyprus will provide the base for any Armenian activity as this would only provoke Turkish military intervention in the Greek portion of the island.

Tensions continue between conservative and radical Arab regimes in North Africa and the Middle East. Added to these are the antagonisms between Shi'ite and Sunni Moslems, and between Moslem fundamentalists and secular regimes. The contest between Morocco and the Algerian- backed Polisario has not ended. Turkey and Greece have not resolved their differences over Cyprus. The growing population of large North African, Middle Eastern, and Southern Mediterranean workers and refugees in Europe has provoked right-wing backlash in several countries; these tensions remain. The influx of pieds noirs from Algeria to Corsica in the 1960s is seen as provoking the Corsican separatist movement. Right-wing terrorist activity directed traditionally against Jews and also Arabs from North Africa has increased in the 1980s. Overlaying these regional conflict sets is the global rivalry between the United States and the Soviet Union.

These antagonisms may provoke open warfare between

nations and will surely continue to lead to subnational conflict that is at the same time international. Occasionally, these subnational conflicts will spill over into the maritime environment, as they have in the past. While it is not apparent at present that any of these will lead to protracted naval guerrilla warfare, they conceivably could lead to incidents of serious consequence.

A CHRONOLOGY OF CONFLICT IN THE MEDITERRANEAN (18)

Palestine Insurgency: 1945-1948

Jewish nationalists fought against local Arabs and the British mandate until achieving independence for the State of Israel on 14 May 1948.

Greek Civil War: 1946-1949

Armed leftist insurgents of ELAS challenged the government. The insurgents were finally defeated in 1949.

First Arab-Israel War: 1948-1949

Israel's existence as an independent state was promptly challenged by the military forces of Egypt, Syria, Jordan, Iraq, and Lebanon. Israeli military forces were successful. The fighting ended with armistice agreements with Egypt, Lebanon, Jordan, and Syria.

Cyprus Emergency: 1952-1959

Cypriot nationalists fought British forces. The fighting ended with the granting of independence to Cyprus.

Algerian Revolutionary War: 1954-1962

Algerian nationalists under FLN leadership fought French forces. The fighting ended with the granting of independence to Algeria.

Suez War: October-November 1956

Israel invaded Egypt in 1956. French and British forces landed to protect the Suez Canal. The fighting ended in a United Nations-sponsored truce.

Spanish-Morocco Border Fighting: 1957-1958

Dispute over the status of Spanish Morocco led to fighting between Spain and Morocco in 1957. The fighting ended with the transfer of Spanish Southern Morocco to Morocco on 10 April 1958.

Subnational Conflict in the Mediterranean Region

Civil War in Lebanon: May-August 1958

Civil war in Lebanon led to intervention by American Marines. The Marines withdrew in October 1958.

Bizerta Crisis: July 1961

A dispute over a French air-naval base led to the armed occupation of Bizerta by French forces. The crisis ended with the transfer of Bizerta to Tunisia in 1962.

Border Fighting between Algeria and Morocco: October 1963

Dispute of a border area led to open fighting between the regular forces of Algeria and Morocco in October 1963. Moroccan forces repulsed Algerian forces. The Organization of African Unity intervened diplomatically to end the dispute.

Cyprus Crisis: December 1963-August 1964

Communal disturbances between the Greek Cypriot and Turkish Cypriot communities led to intervention by both Greek and Turkish armed forces. Ultimately, the United Nations dispatched a peacekeeping force to supervise the ceasefire.

Syrian Coup: February 1966

After bloody fighting, rebel factions in the Syrian armed forces overthrew the civilian government.

Lake Tiberias Incident: 7 April 1967

A dispute over the demarcation line on Lake Tiberias led to an artillery exchange between Israel and Syria leading to aerial combat.

Six-Day War: 5-10 June 1967

Following armed provocation by Jordan and Syria, Israel invaded Egypt, Syria, and Jordan. The brief war ended with Israeli occupation of vast territory.

War of Attrition: 26 October 1968 - 7 August 1970

Egyptian military units opened fire at Israeli positions on the east bank of the Suez Canal provoking Israel to launch a series of commando raids against Egypt. Artillery barrages and raids by Egyptian and Israeli commandos continued until July 1969 when the Israeli air force began raids on Egyptian targets along the Canal, and later against military targets deep inside Egypt. Egypt and Israel agreed to a ceasefire on 7 August 1970.

Civil War in Jordan: September 1970

In the late 1960s, Jordan permitted its territory to be used as a major base for Palestinian guerrillas, but by 1970, Palestinian strength had grown to 20,000 armed guerrillas who were openly challenging King Hussein's authority. Armed clashes between the Palestinians and the Jordanian army escalated into open warfare in September. The civil war officially began in September 17 when Jordanian units moved to clear Amman of all Palestinian guerrillas. The Palestinians were pushed back to an area on the Syrian border where they remained until July 1971 when the Jordanians drove them out of the country.

Israeli Air and Ground Engagements with Lebanon, Syria and Egypt: February-November 1972

In response to terrorist attacks on Israeli borders and Israeli targets abroad, Israeli air and ground forces attacked Palestinian guerrilla bases in Syria on February 24, March 1, and September 8; in Lebanon on February 27, June 23, and September 16; on June 13, Israeli and Egyptian air forces engage in a brief battle; after three weeks of sporadic clashes, heavy fighting broke out between Israel and Syria on November 21 and continued sporadically into December.

Attempted Coup in Morocco: 16 August 1972

Rebel aircraft attacked King Hassan's aircraft and Palace in an unsuccessful coup.

Israeli Reprisal Raids in Lebanon and Syria: January-September 1973

In response to Palestinian terrorist attacks in Israel and against Israeli targets abroad, Israeli jets and Israeli commandos struck targets in Syria on January 8 and September 13 and in Lebanon on February 21 and April 9.

Subnational Conflict in the Mediterranean Region

Guerrilla Invasion of Lebanon: 9 May 1973

Guerrillas with Syrian support invaded Lebanon on May 9. Lebanese armed forces pushed the invaders back to the Syrian border on May 11.

Yom Kippur War: 6-24 October 1973

Eygptian and Syrian forces attacked Israel on 6 October 1973. Israeli forces counterattacked on both fronts and on October 15, crossed the Suez Canal. A ceasefire was arranged on 24 October 1973. A 7,000-man United Nations peacekeeping force was sent to the Sinai on October 27.

Israeli Reprisal Raids in Lebanon: May 1974-August 1975

In response to continuing Palestinian attacks on Israeli targets in Israel and abroad, Israeli aircraft and commandos attacked targets in Lebanon on May 16, 17, 19 and 21, June 18-20, October 31, November 1 and 14, 1974; and January 12-16, May 25, July 4-7, August 5, and August 21-22, 1975.

Turkey's Invasion of Cyprus: July 1974

Following the overthrow of President Makarios of Cyprus by EOKA-B terrorists and Greek officers in the Cyprus National Guard, Turkey invaded Cyprus on July 20 leading to the de facto partition of the island into Greek and Turkish sections.

Civil War in Lebanon: 1975-1976

Tensions between Christians and Moslems, exacerbated by the growing strength of Palestinian guerrilla groups in southern Lebanon, erupted into civil war in 1975. The Lebanese army virtually disintegrated as fighting continued between private militias. Continued conflict led to Syrian intervention on 31 May 1976. Syrian forces entered Beirut on 10 November 1976. The Syrian pulled back to positions in northern Lebanon during Israel's 1982 invasion where they remained at the end of 1982.

Fighting in the Western Sahara: 1975

The withdrawal of Spain from its western African colony, Spanish Sahara, set off a war between Morocco and the

Algerian-backed guerrillas of <u>Polisario</u>. Occasionally Algerian and Moroccan forces clashed directly. Spain's original plan was to divide the territory between Morocco and Mauritania, which <u>Polisario</u> opposed. Moroccan forces occupied the capital of Spanish Sahara in December 1975. Mauritania renounced all territorial claims in August 1979. The fighting between Moroccan forces and <u>Polisario</u> guerrillas continues.

Egypt-Libya Border War: 21-24 July 1977

Strained relations between Egypt and Libya led to brief clashes involving tanks and aircraft on the Egyptian-Libyan border on 21 July 1977. Egyptian President Sadat ordered a ceasefire on July 24.

Egyptian Commando Raid: February 1978

In an effort to rescue hostages held aboard an airliner hijacked by Palestinian terrorists, Egyptian commandos landed at Larnaca Airport in Cyprus. They were met by Cypriot security forces and a brief gunbattle ensued. All of the Egyptians surrendered to Cypriot authorities and later were returned to Egypt.

Israeli Invasion of Lebanon: 14 March-13 June 1978

Provoked by an Al Fatah attack on a bus outside Tel Aviv in which 37 persons were killed, Israel invaded southern Lebanon on 14 March 1978. After United Nations forces were sent to maintain peace in southern Lebanon, Israel withdrew on June 13.

Libyan Invasion of Chad: April 1979

For years, Libya had supported rebels in northern Chad against the French-backed government forces. In April 1979, a small Libyan force invaded the uranium and iron-rich Aouzou strip. They were repulsed by Chad's regular army.

Commando Raid on Tunisia: 27 January 1980

A group of approximately 50 commandos, mostly Tunisian expatriate workers in Libya, launched a coordinated attack on the army barracks and police station in the Tunisian border town of Gafsa. The attack was repelled and most of the attackers captured, 41 persons were killed. Tunisia accused Libya of organizing the raid.

In a gesture of support for the Tunisian government, France dispatched a small rural convoy to show the flag off the Tunisian coast. On February 4, demonstrators in Tripoli stormed and sacked the French Embassy. France charged that the attack on its embassy and its consulate in Benghazi were government organized.

Uprising in Syria: 4-9 March 1980

Unrest attributed to the Moslem Brotherhood led to antigovernment riots in Aleppo and Hama which turned into a bloody confrontation between armed opponents of Syrian President Assad and 10,000 Syrian troops. The brief revolt was suppressed by March 9. (On 17 March 1981, Syrian gunmen attempted to assassinate the exiled leaders of the Moslem Brotherhood in West Germany.)

Second Libyan Intervention in Chad: June 1980-November 1981

Libyan troops again entered Chad in June 1980 to intervene in Chad's civil war and press Libya's territorial claims. Libyan troops participated in the fighting until 12 December 1980 when they entered Chad's capital on behalf of Chad's president. The Libyan forces remained in Chad until November 1981.

Aerial Clash Over Gulf of Sidra: 19 August 1981

The United States has regularly resisted Libyan territorial claims to the Gulf of Sidra by holding naval manouevers in international waters off the Libyan coast. American F-14 fighters shot down two Libyan fighters that fired at them approximately 60 miles from the Libyan coast.

Second Israeli Invasion of Lebanon: 5 June 1982

In response to continued Palestinian terrorist attacks on Israeli targets abroad and continued Palestinian shelling of Israeli settlements near the Lebanese border, Israeli forces invaded Lebanon in strength on 5 June 1982. (There had been several Israeli invasions since 1978.) Israeli units pushed Palestinian units back, then surrounded Beirut which was subjected to artillery and aerial bombardment. The Palestine Liberation Organization was permitted to withdraw

its remaining forces in Beirut in September under international supervision by Italian, French, and American
forces. At the end of 1982, Israeli forces continued to
occupy southern Lebanon up to Beirut.

NOTES

1. These figures are derived from a chronology of international terrorism maintained by The Rand Corporation.
Figures maintained by the United States Government differ in
the total number of terrorist incidents worldwide, but show
a similar distribution pattern among countries.

2. The activities of the Irgun and Stern Gang are de-
scribed by J. Bowyer Bell, Terror Out of Zion, Irgun Zva;
Leumi; LEHI, and the Palestine Undergound, 1929-1949, St.
Martin's Press, New York, 1977. See also Menachim Begin, The
Revolt: The Story of the Irgun, Hadar Publishing Company,
Tel Aviv, 1964.

3. For a description of the EOKA-B, see Charles Foley
and W.J. Scobre, The Struggle for Cyprus, Hoover Institution
Press, Stanford, 1975.

4. The activities of the FLN are described by Martha
Crenshaw Hutchinson, Revolutionary Terrorism: The FLN in
Algeria, 1954-1962, Hoover Institution Press, Stanford,
1978. The activities of the OAS and barbouzes are described
by Paul Henissart, Wolves in the City: The Death of French
Algeria, Simon and Schuster, New York, 1970.

5. Frantz Fanon, The Wretched of the Earth, Groove
Press, New York, 1963.

6. The FLN's influence on Fatah is discussed in John
W. Amos II, Palestinian Resistance: Organization of a Natio-
nalist, Pergamon Press, New York, 1980, p. 49; Edgar O'Bal-
lance, Arab Guerrilla Power: 1967-1972, Faber and Faber,
London, 1974, pp. 23, 26, 49-50; Zeev Schiff and Raphael
Rothstein, Fedayeen: Guerrillas Against Israel, David McKay
Company, Inc., New York, 1972, p. 8.

7. It is difficult to discuss Palestinian terrorism
with Arabs understandably sympathetic to the Palestinian
cause. The discussion inevitably becomes adversarial. Indeed, some see the use of the term "terrorism", which is a
pejorative, as simply the latest in a series of deliberate
slurs calculated to distort the Western image of the Arab:
The Arab as terrorist has replaced the Arab as lascivious
keeper of harems. Some argue that the tactics employed by
the Palestinians are justified since the Palestinians are
the victims of greater terror by the Israelis and that
terrorist tactics are the only means of struggle available
to them - the traditional defence of terrorism. Some Palestinians assert that Palestinian terrorism is, in fact, not

Palestinian at all, but rather comprises actions carried out by gunmen in the employ of unscrupulous Arab governments. Yet the fact of Palestinian terrorism remains, and in a sense has been successful. That there is now pressure for an Israeli withdrawal and the creation of a Palestinian home-land, that the Palestinian Liberation Organization may now be accorded international recognition as the legitimate government of a stateless people, is owing at least in part to the success of Palestinian terrorists in bringing their cause violently and dramatically before the eyes of the world. Without endorsing terrorism, one must wonder what success they could have won had they operated within the established bounds of conventional warfare and polite diplomacy.

8. Andres Cassinello Perez, "ETA y el problema vasco", an unpublished manuscript presented at the Seminario sobre Terrorismo Internacional, held in Madrid, 10-12 June 1982.

9. Claire Sterling, The Terror Network, Holt, Rinehart and Winston, New York, 1981, pp. 195-196. Sterling cites "Western intelligence services" for this item. Although I have heard similar reports concerning Algerian support for European separatist movements, like most information about the international connections, it must be viewed with caution.

10. The Libyan involvement in various plots are des-cribed by John K. Cooley, Libyan Sandstorm, Holt, Rinehart and Winston, New York, 1982.

11. The episode is described by Robert Katz, Days of Wrath: The Ordeal of Aldo Moro, Doubleday & Company, Inc., Garden City, New York, 1980, pp. 187-189.

12. Complaints against the French position are found in Cassinello, op. cit. They were also voiced at the Council of Europe's Conference on the Defence of Democracy Against Terrorism in Europe: Tasks and Problems, Strasbourg, 12-14 November, 1980.

13. Gerard de Villiers, "The Terrorist Plan for 1982", Paris Match, 16 April 1982, pp. 48-51.

14. See, for example, Robert Charm, "Terrorists See Offshore as Tempting Target", Offshore, January 1983; Robert W. Denton, "Protection of Offshore Energy Assets", Naval Engineers Journal, December 1976, pp. 87-91; John F. Ebersole, "International Terrorism and the Defense of Off-shore Facilities", U.S. Naval Institute Proceedings, September 1975, pp. 54-61; ISIS Associates, Incorporated,

International Symposium on Maritime Security & Terrorism, Arlington, Virginia, 21-23 September 1981; Merle MacBain, "Will Terrorists Go to Sea?", Sea Power, January 1980, pp. 15-24; Douglas G. Macnair, "Terrorism in the Maritime Environment", Terrorism and Beyond, The Rand Corporation, R-2714, December 1982, pp. 273-275, and Douglas G. Macnair, "'The Nature of the Beast', A Soliloquy on Maritime Fraud, Piracy, and Terrorism", Journal of Security Administration, June 1982, pp. 41-47.

15. Offshore, "Worldwide Exploration and Production: Prospects Bright off Mediterranean", 20 June 1982.

16. Tim A. Cornitius, "Submersible Fleet Growth Surpasses Early Expectations", Offshore, August 1982.

17. Council of Europe, Parliamentary Assembly, Political Affairs Committee and Legal Affairs Committee: Subcommittee on Terrorism in Europe, "Summary Report of the Conference on The Defense of Democracy Against Terrorism in Europe: Tasks and Problems", Strasbourg, 9 January 1981.

18. This chronology is based in part (1945-1967) on the excellent chronology compiled by David Wood in Conflict in the Twentieth Century, Adelphi Papers, No. 48, The Institute for Strategic Studies, London, 1968.

Chapter Seven

THE MILITARY PRESENCE OF THE RIPARIAN COUNTRIES

Maurizio Cremasco

THE FRAME OF REFERENCE

It almost goes without saying that the Mediterranean
is an area abounding in situations of potential instability,
latent crises and endemic conflicts. Just a quick survey of
the area and an extremely schematic political analysis of
the situation will in fact suffice to confirm this observation.

Yugoslavia is faced not only with a difficult economic
conjuncture but also with the thorny question of Kosovo
demands for greater autonomy, a problem which involves more
than just ethnic minorities and greater political representa-
tion.

Albania, locked in virtual international isolation,
will soon have to deal with the delicate problems posed by
the post-Hoxa period. The new regime may gradually adopt a
different foreign policy. But in any case, a factor which
will undoubtedly affect future Albanian international
choices is the need to restore credibility to the country's
armed forces, currently at a low level of operational readiness. (1) If Albania reopens its doors to the Soviet Union
(Tirana's major arms supplier until the 1958 schism) and
eventually restores its former political and military ties
with the leader of the Communist bloc, this would have
serious repercussions not only on the Balkan region but on
the entire Mediterranean area. However, this is still an
unlikely hypothesis. Even after secretary Brezhnev's death,
Tirana's reaction to the new Soviet leadership was very cool.

In the Aegean, the disputes between Greece and Turkey
over Cyprus, division and control of air space, the limits
of their respective territorial waters, sovereignty over the

continental shelf and rights to exploit the seabed are still smoldering. The two countries' internal situations do not seem to make the search for agreements and complete normalisation of relations any easier. In the wake of Papandreou's electoral victory, Greece has adopted a more explicitly nationalistic attitude towards the problem of relations with Ankara, rekindling old controversies and setting in motion mechanisms of confrontation which may lead to a new crisis. Turkey, on the other hand, must deal not only with the problems posed by a difficult economic situation but also with the need to establish a domestic political order which would guarantee, with the reinstatement of democracy, the country's governability.

The Middle East represents the most difficult political and military knot to untie. Elements of instability are present throughout the region and have their origins in a number of different factors and circumstances; the repercussions of Israel's foreign and military policy and the unresolved Palestinian question; the unstable domestic situation of many countries; the two superpowers' interests in the area, the role they play in affecting the outcome of the recurring crises, and their attempts to arrive at negotiated solutions to the crises; inter-Arab rivalry and repercussions of events in the Persian Gulf.

In the Maghreb, the problem of the ex-Spanish Sahara remains open. The Polisario Front, supported by Algeria, continues its guerrilla activities while the plan for self-determination of the Saharoui people elaborated by the Committee of Seven of the Organization of African Unity has failed as an attempt to promote a political solution to the crisis.

Finally, the question of sovereignty over Gibraltar, which opposes two NATO members, Spain and Great Britain, has still not been resolved.

What emerges from the above frame of reference is that the Mediterranean is an area divided into a number of different "tension zones". This fact makes it unrealistic, if not impossible, to consider the area as a single entity to which a common parameter of political and strategic analysis can be applied. The tensions in the various zones derive, in fact, from problems which are very diverse in terms of historical and ethnic roots, political and economic interests, and security needs.

This fragmentation into a number of distinct "tension

zones" does not, however, exclude the possibility, should the tension in one zone break out into open conflict, that other zones of the Mediterranean or countries belonging to another region may be affected or that the two superpowers and East-West relations may be directly or indirectly involved.

Also of significance is the fact that the tension zones are not located along the borders between NATO and Warsaw Pact countries and that the situations of latent crises cannot be attributed to elements of confrontation between the two alliances in southern Europe. Instead, they exist either within one of the alliances (the Greek-Turkish dispute; Romania's often eccentric foreign policy with respect to the line set by the Soviet Union) or outside NATO's area of responsibility, in zones not covered by the Treaty.

Finally, the trend in the Mediterranean is likely to be towards a multiplication of the tension zones (or towards an accentuation of the existing crises) as the result of two concomitant developments: the integral application of the Law of the Sea, with the institution of exclusive economic zones (EEZs) up to 200 miles off the shores of the coastal countries and the extension of territorial waters from 6 to 12 miles, on the one hand, and, on the other, progress in mining technology which will make it possible and economically feasible to exploit seabed resources.

The problem in the Mediterranean appears particularly complex. Given the Mediterranean's geography, the institution of exclusive economic zones will inevitably lead to a series of superimpositions and hence to motives of controversy. In fact, it is impossible to draw a line 200 miles from the coast of any Mediterranean country without it cutting across and overlapping the corresponding line traced from the coast of another country (in some cases island territories or part of the continental territory of another country may even be included). Even the extension of territorial waters does not appear of easy application. The extension of the territorial waters of the Greek islands from six to twelve miles would increase Athen's sovereignty over the Aegean Sea from 35% to 64%; would reduce the extent of international waters to 25%; and would virtually deprive the Turkish ports of the Aegean of direct outlets to international waters. (2)

Progress in mining technology, which will make it possible to intensify exploration for and development of underwater resources - as soon as technology reaches the point where it is economically feasible to carry out exploration and extraction activities at depths of up to 1000 metres, the total exploitable area will pass from the current 15% to 22% (3) - will just as inevitably tend to make it more difficult for the countries whose EEZs overlap to reach agreements on their exploitation.

Not only are the politico-economic controversies linked with the determination of sovereignty rights likely to increase in the future - the disputes between Greece and Turkey regarding the limits of sovereignty over the Aegean continental shelf and between Malta and Libya over the shelf in the central Mediterranean are still unresolved - but so are the contrasts regarding the rights of free passage and freedom of navigation.

This eventuality as a source of crises and conflicts should not be underestimated. A look at history, even recent history, should make this clear. In 1967, Egypt closed the Strait of Tiran, making it impossible for ships directed towards Israel to enter the Gulf of Aqaba. Twenty-four hours later Israel declared war. The threatened closure of the Strait of Hormuz in 1979 because of the Iran-Iraq conflict was the principal motive behind the American decision to send a fleet to the Arabian Sea. The same decision was taken by two European countries, France and Great Britain. In August 1981, F-14 fighters of the U.S. Sixth Fleet carrying out manoeuvres in the central Mediterranean clashed with Libyan SU-22 planes in the sky above the Gulf of Sidra, in a zone 60 miles from the coast that Tripoli had unilaterally included in its territorial waters. The shooting down of the two Libyan planes that attacked the Tomcats of the aircraft carrier Nimitz was a clear indication of the importance the United States attributes to the problem of freedom of navigation and of its resolve not to recognise or submit to other limits beyond those established by the international laws currently in force.

On the other hand, the definition of "innocent passage" and the distinction between warships and merchant ships tend to get blurred in times of crises: when the motives of having a naval presence to "show the flag" and signal interest in the course of events or exert political pressure are pre-eminent; or when merchant ships are used as the principal means of transporting arms and supplies to one of the warring parties.

The Military Presence of the Riparian Countries

In the Mediterranean, naval forces have often been used by both the superpowers and the coastal countries. (4) They have been employed as a foreign policy instrument to send signals or warnings. For instance: in 1946, the battleship Missouri, flanked by the light cruiser Providence and the destroyer Power, paid a visit at the port of Istanbul to signal to the Soviet Union the importance the United States attributed to the security of Turkey; in October 1967, Soviet ships visited Port Said to deter Israel's reactions after the sinking of the destroyer Eilat; in October 1973, three aircraft carrier task groups of the U.S. Sixth Fleet were present to the south of Crete. Naval forces have also been used as instruments of war. For example: the landing of the U.S. Marines in Lebanon in 1958; the sinking of the Israeli destroyer Eilat by fast attack craft during the 1967 conflict; the clashes between the Israeli and Egyptian navies during the 1973 conflict; the bombing of the Lebanese coast and Beirut by Tel Aviv's naval units during the summer 1982 crisis. And also as the principal means of moving men and material: naval units were used to send to Syria the Moroccan contingent which fought on the Golan front in October 1973; during the Arab-Israeli war of Yom Kippur, about 85% of Soviet military supplies to the Arab countries and over 70% of U.S. supplies to Israel were carried by ship. Air forces are used for much the same purposes, but rarely as a foreign policy instrument. Their low degree of suitability as an instrument of exerting pressure derives from the fact that, although they can be used selectively, they cannot be graduated and cannot constitute a fixed presence. Israel has been able to repeatedly use combat aircraft flying over its adversaries' principal cities at very low altitude and very high speed as an instrument of pressure and intimidation in the typical no-war no-peace situation thanks to the contiguity of enemy-countries' territory and their lack of an effective air defence system.

The prospect of air and naval forces actually being used has taken on new dimensions in recent years, especially in the Mediterranean, owing to the development of new technologies.

Radar planes (such as the AWACS or the E-2C Hawkeye) can supply precise information on the movements of fleets and on the consistency of an eventual air threat; they can operate as command and control centres; they can direct the attacks of interceptors and fighter-bombers.

The Military Presence of the Riparian Countries

The new generations of fighter planes are endowed with high cruising and attack speeds; they are equipped with advanced navigation and firing systems, have a long radius of action and an elevated war load capacity; they can be armed with sophisticated weapons ("smart" bombs, air-to-surface missiles with electro-optical or radar terminal guidance or "fire and forget" capabilities) and fitted with electronic warfare systems.

Modern warships are equipped with anti-aircraft and anti-missile defence systems which are radar-controlled and completely automatic in their loading and firing operations; with offensive systems consisting in surface-to-surface missiles; and with sophisticated ASW systems (variable depth sonars, anti-submarine missiles and ASW helicopters).

The biggest revolution has been in the field of missiles. Anti-ship (air-to-surface and surface-to-surface) and anti-aircraft missiles are today the most lethal weapons of air-sea warfare. Because the anti-ship missiles can be mounted even on low displacement ships such as corvettes, fast patrol vessels, and hydrofoils, or can be deployed on ground on mobile bases, they have become the principal weapon of all the maritime forces that operate in the Mediterranean.

Paradoxically, however, it is precisely these techno-logical developments which place certain limits on the use of naval forces in the Mediterranean because of the sea's particular geographic conformation.

The Mediterranean is a relatively large sea (2,511,000 sq.km.), enclosed by the Turkish Straits to the east and the Strait of Gibraltar to the west, and divided into compart-ments by a number of choke points: the Sicilian Channel, which practically divides the sea into two distinct basins; the Otranto Channel; the entrances to the Aegean Sea to the east and west of the Island of Crete; the passages between the more than 3000 islands of this sea. The east-west dimensions of the sea are appreciable, while the north-south dimensions are very modest. The length of the Mediterranean from Gibraltar to the Turkish coast is just over 4000 km. The maximum width, from the Gulf of Trieste to the African coast of Sidra is only about 1800 km. Other distances vary from 1 to 3 km. at the Turkish straits to 13 km. at the Strait of Gibraltar; from 145 km. at the Sicilian Channel to some 450 km. from Cape Passero to Benghazi; from just over 300 km. from Crete to Tobruk to about 550 km. from the

Turkish to the Egyptian coast; from some 250 km.   from   Cape
Teulada in Sardinia to Annaba in Algeria to over 600 km.
from Toulon to the Algerian coast.

This means that any point on the Mediterranean – and
in some cases even more or less vast parts of the territory
of the coastal countries – can be covered, given their
radius of action, by the modern land-based fighter-bombers
and, given their high operational speeds (from 360 to 450
knots), in a relatively short time. The navigation systems
they are equipped with permit accurate arrival on their targets, even if mobile, and the missiles they can be armed
with assure a high kill probability.

Moreover, the choke points in the Mediterranean constitute obligatory points of passage and can easily be
controlled, made difficult to transit through, or completely
blocked with the use of attack submarines, fast missile
units, or mines. Even the Sicilian Channel, notwithstanding
its width, has a number of relatively shallow points because
of the wide extension of the European and African continental shelves which limit full freedom of transit and
manoeuvres of submarines.

In addition, the high thermal gradient, especially in
the summer season, the elevated salinity of the sea, (5) the
uneven conformation of the seabed, and the heavy traffic
(daily, more than 3000 ships of over 1000 tons and about
5000 other boats of various shapes and sizes), make locating
submarines, and hence submarine warfare, especially difficult.

Finally, if mobile land-based anti-ship missiles are
deployed near the zones with the greatest amount of traffic
or through which transit is obligatory, this would represent
a further significant threat which is certain to have an
impact on the operations of naval forces (particularly
amphibious operations) and on maritime traffic. In the near
future mobile anti-ship missiles which can be launched from
the ground and with a range much superior to the current
100–120 km. will be available on the international arms
market at relatively low prices. Their proliferation in the
military arsenals of the Mediterranean countries would
further limit the use of naval forces near the coasts and
would further reduce operational maritime space in the Mediterranan.

In conclusion, it can be said that the Mediterranean
has today become a much smaller sea in geostrategic terms.

The Military Presence of the Riparian Countries

The vulnerability of ships has increased, especially if they do not have anti-missile defences or if they are not operating under an anti-air umbrella. The role of land-based attack air forces has grown. The threat to maritime traffic has become more penetrating and diversified, especially near the choke points. The use of naval forces as a foreign policy instrument has become more difficult and risky. All this against the backdrop of a situation already riddled with instability and unresolved political disputes and in which there is the prospect of increased conflict due to application of the Law of the Sea treaty.

A major factor determining this geostrategic shrinking of the Mediterranean and the deep political and military transformations in the area is the qualitative and quantitative build-up of the air and naval forces of the coastal countries.

This build-up of forces in the Mediterranean raises a number of critical questions.

To what extent will this build-up affect each country's security perceptions and will it lead to a further spiralling of the race for more and increasingly sophisticated arms in the Mediterranean area?

To what degree might the build-up itself become an element which generates crises, encouraging the use of military force to resolve eventual disputes?

To what extent might it constitute a real threat to maritime traffic and naval operations and might it become the means with which some countries attempt to unilaterally impose their sovereignty on certain zones of the sea, limiting freedom of navigation?

To what degree might it affect the freedom of action of the superpowers' air and naval forces in the Mediterranean and to what extent might it represent an obstacle to the use of military power as an instrument of pressure and intimidation? (6)

To what degree and in what way will the build-up influence the search for agreements or measures to control the proliferation of arms in the Mediterranean area?

A SUMMARY DESCRIPTION OF THE CURRENT SITUATION

From a summary analysis of the developments in the air and naval forces of the coastal countries from 1970 to 1982, a number of remarks and considerations can be put forward. (7)

- The air and naval forces of the Mediterranean countries (especially those of the Arab countries of North Africa and the Middle East) have been significantly enhanced both in numbers and in quality. Not only have arms and equipment been purchased in greater quantities; the purchases have also led to greater diversification of capabilities through the procurement of new high performance weapons systems with advanced characteristics.

- The air forces have normally benefited more than the naval forces. Today, all the Mediterranean countries, with the exception of Albania (100), Morocco (97), Spain (210), and Tunisia (8), have air forces with more than 300 combat aircraft (thus excluding transport and training aircraft and helicopters). Except in a few cases (Albania, Syria, Spain), the total number of fighter-bomber squadrons exceeds that of the interceptor squadrons which means that attack capability has been substantially privileged with respect to defence capability. This may be due in part to the fact that the countries concerned have adopted an air doctrine which concentrates on superiority by means of counter-aviation rather than through air combat; in part to the awareness of the importance of the land front for the country's defence and hence the need to give direct support to the ground forces; in part for political reasons since offensive air forces are one of the most visible "status symbols" of military strength. Finally, for the Arab countries (but not only for them) we must not forget how the use of Israeli air forces during the 1967 war influenced their arms procurement programmes.

The "status symbol" motive and the fact that it is best served by the air forces because of their high technological content has also played a role in the decision to give priority to the development of the air force rather than the navy. An equally important factor which has determined such choices is the repercussions of the so-called "mirror effect", by which the arms purchases of one country tend to be matched by neighbouring countries and, more generally, by all the countries in the region.

- The build-up of the air forces in terms of quality has been striking in many respects. The air forces of the Mediterranean countries now deploy medium bombers (not only the old Egyptian Tu-16 Badgers, but also the more modern

214

Libyan Tu-22 Blinders) and a whole series of new generation combat aircraft: F-15, F-16 and Kfir C-2 (Israel); F-16 (Egypt); MiG-23-27 (Algeria, Libya, Syria); MiG-25 (Libya, Syria); Su-20 (Algeria, Libya, Syria); Mirage F-1 (France, Greece, Libya, Morocco, Spain). In addition, the older but still valid F-4 (Egypt, Greece, Israel, Turkey); F-104 (Greece, Italy, Turkey) (8); Mirage III and Mirage V (Egypt, France, Israel, Libya, Spain).

The weapons systems include air-to-air missiles: Soviet AA-2 Atoll, U.S. Sidewinder and Sparrow, French Falcon and R-550 Magic, Israeli Shafir; and air-to-surface missiles: Soviet AS-1 Kennel and AS-5 Kelt, French AS-37 Martel and AM-39 Exocet, U.S. Bullpup, Maverick and Harpoon, German Kormoran.

Of course the sophistication of the means is not always matched by effective operational capacity. In many countries the weapons systems cannot be used to their full potential because the training of crews and pilots is insufficient. Inadequate technical support and maintenance lower the efficiency of the systems, while lack of sufficient logistical support means that protracted war operations could not be sustained.

But in any case, the air threat in the Mediterranean has definitely grown in terms of greater coverage, high intervention speed, more diffused presence of modern weapons systems endowed with enhanced firing accuracy and high destruction potential.

- For the naval forces, the situation can be summarised as follows:

a) Greater diffusion of submarines. Greece, Yugoslavia, Libya, Spain and Turkey have all increased the number of their submarines. Libya, which in 1970 had no submarines, now has five of the Foxtrot class supplied by the Soviet Union. (9) Algeria, Morocco, Tunisia and Syria still do not have any submarines. Submarines continue to be a typical component only of the European navies, and of the Israeli, Egyptian and Libyan navies. But the situation might change if the submarine together with missile-armed ships comes to be considered a more suitable means for enforcing a "sea denial" strategy, for instituting naval blockades or for threatening in a more invisible, and hence more insidious form, transit in those zones of the sea in which a country wants to impose its sovereignty. It is difficult to say whether the sinking of the General Belgrano cruiser in the recent war for the Falkland Islands and the role played by the British nuclear submarines in confining the Argentine fleet to its ports will encourage the Mediterranean countries that still are not equipped with submarines to reassess their importance. However, for many of them, the real defence needs to justify such a procurement would be lacking.

b) Increase in the number of ex-novo procurements of frigates. The multi-purpose frigate – a ship equipped with diverse and highly sophisticated weaponry and, in the case of the new generations, capable of carrying ASW helicopters – is becoming a more and more typical component of the navies of all the Mediterranean countries. Over the past ten years frigates have been purchased by the Algerian Navy (Soviet Koni-class), the Libyan Navy (English Vosper Thornycroft Mark-7 frigate, the Syrian Navy (Soviet Petya-class), the Tunisian Navy (U.S. Savage-class) and the Moroccan Navy (Spanish Descubierta-class). Other navies have purchased new generation frigates with enhanced anti-sub, anti-air and anti-ship capabilities (Italy now has Lupo and Maestrale-class, Spain Descubierta-class, Greece Kortenaer-class and Turkey Berk-class frigates).

c) A noteworthy increase in light missile units (corvettes, fast attack craft and hydrofoils). This is the most evident and strategically significant development in the area.

Almost all the Mediterranean countries are today equipped with fast attack craft (FAC) armed with missiles. For many, almost all the Arab countries and Israel, they constitute the principal offensive component of their naval forces. The types of units, divided according to class and country are the following:

- Corvettes: Nanuchka-class armed with SS-N-2C missiles (Algeria and Libya); Wadi M'ragh and Dat Assawari-class armed with Otomat missiles (Libya); Aliya-class armed with Gabriel missiles (Israel).

- Fast attack craft: OSA-I-class and OSA-II-class armed with SS-N-2 Styx missiles (Algeria, Yugoslavia, Libya, Egypt and Syria); Hoku-class armed with SS-N-2 missiles (Albania); La-Combattante-II and La-Combattante-III-class armed with Otomat, Exocet and Penguin missiles (France, Greece, Libya and Tunisia); Reshef and Saar-class armed with a Harpoon and Gabriel missiles (Israel); Kormoran-class armed with Exocet missiles (Morocco); October 6th and

Ramadan-class armed with Otomat missiles (Egypt); Dogan and Kartal-class armed with Harpoon and Penguin missiles (Turkey).

- Hydrofoils: Sparviero-class armed with Otomat missiles (Italy); Flagstaff-class armed with with Gabriel and Harpoon missiles (Israel).

These units are relatively inexpensive and have a particularly favourable cost-benefit ratio in terms of capacity to undertake various missions. The FAC are very fast - with maximum speeds of about 35 knots on the average; faster, up to 50 knots, for the hydrofoils - extremely manoeuvrable (but suffer in rough sea conditions), and are armed with anti-ship missiles and at least one small or medium-calibre cannon for anti-air and anti-helicopter defence. They are therefore in a position to undertake various tasks: patrol and control (against the infiltration of saboteurs or landing attempts by commando units); surveillance of the EEZs, interruption or harassment of commercial maritime traffic; attacks on naval formations with "wolf pack", ambush or hit-and-run tactics.

The anti-ship missiles with which they are equipped and their high speeds make them a dangerous threat even for very large and better armed warships.

However, the FAC appear particularly vulnerable to air attacks and will probably operate in zones not too far from the coast, even if their range of action would make it possible for them to cover vaster areas, (10) and in relatively calm sea conditions.

The FAC therefore appear particularly suited to operations around the straits, in the sea areas where there are a number of obligatory passageways, in the gulfs, around the choke points of the Mediterranean, and to patrol the EEZs.

d) Another development of the naval forces which deserves mention is the growth of the amphibious capabilities of the Libyan Navy. Libya has ordered nine C-107-class LCTs (landing craft, tank) from Turkey and has already received two. The nine LCTs together with the two 2800-ton LSTs (landing ship, tank), the three Polish-made, Polnochny-class LSMs (landing ship, medium) and the 3100-ton roll-on/off ship will bring the Libyan landing forces to a level, at least in terms of the means available, superior to that of the amphibious forces of the other North African countries (Algeria, Morocco and Tunisia) and inferior only to Egypt's amphibious forces.

The Military Presence of the Riparian Countries

One last consideration which emerges from an analysis of the situation in the Mediterranean concerns the roles played by the two superpowers and by the European countries as suppliers of arms. The United States supplies the NATO countries of the Mediterranean, Egypt (after Cairo broke off relations with Moscow), Morocco and Tunisia. The Soviet Union supplies Yugoslavia, Egypt (until 1974), Algeria, Libya and Syria with arms.

In recent years, political considerations - the need to avoid exclusive dependence on one of the superpowers for the country's military needs and thus limit the superpowers' possibilities of using this dependence to exert diplomatic pressure on the country in times of crisis - motivated many countries to diversify their sources of weapons, even though they knew that this decision would involve difficult technical, logistic and training problems.

The European countries were the only alternative, not only because they are in a position to supply arms and equipment at a technological level comparable (in some cases superior) to that of U.S. and Soviet products, but also because the supplies were not dependent on political concessions in return (for example, access to air or naval infrastructures).

With regards, once again, only to the air and naval forces: France has sold arms systems to Egypt, Libya, Morocco and Tunisia; Great Britain to Egypt, Algeria, (11) Libya, Morocco and Tunisia; Italy to Libya, Morocco, Egypt and Tunisia; while Turkey has recently begun to build naval units for Libya.

Even this very sketchy picture of the air and naval forces of the countries bordering on the Mediterranean makes it evident that the concentration of military capability in the Mediterranean basin is indeed impressive. The picture becomes even more complex and alarming if we also consider the presence of U.S. and Soviet air and naval forces and the evident trend towards further militarisation which emerges from an examination of the Mediterranean countries' plans to build up their armed forces, the United States' military aid programmes and the Soviet Union's new commitments to supply arms.

The Military Presence of the Riparian Countries

THE IMPLICATIONS OF THE NEW SITUATION

An attempt to single out the implications of the new situation amounts in substance to an attempt to answer the questions posed in the first section of this paper.

The security choices of a country are never the result of consideration of just one parameter. Other factors besides defence needs (that is, besides an assessment of current and impending threats) often come into play. These may include the political role the country aspires to play in the regional context; the leadership ambitions that a strong military force may nurture; the international ties with the politics of one or the other superpower (but only insofar as a substantial convergence of interests exists); the means to dedicate a considerable percentage of the state budget to military spending (characteristic of the oil-exporting countries); the intention to develop the arms industry as a means of reducing dependence on other nations and as part of a broader scheme for the industrialisation of the country, counting on the transfer of technology and know-how from the military to the civilian sector.

The primary factor motivating a build-up of military capabilities remains, however, the perceived threat to the country's security which, in turn, is a function of the elements of regional instability and of the persistence of problems and contrasts with neighbouring countries. It was thus that the Arab-Israeli conflict decisively spurred the enormous growth of military arsenals in the Middle East, just as the ex-Spanish problem is driving Morocco to build up its armed forces.

Moreover, a build-up of military strength in one country is perceived by neighbouring countries as an increased threat to their own security, setting off that vicious circle of action-reaction typical of the arms race.

For the European countries of the Mediterranean the problem is even more complex. Their defence policy choices are usually a part of the collective decisions taken by NATO in response to the Warsaw Pact's increases in military potential. In the Mediterranean theatre, the Warsaw Pact's potential was increased mainly through a build-up of Soviet air and naval forces (the Soviet Mediterranean fleet was increased in numbers and quality; the Soviet Naval Aviation deployed Backfire bombers at its Crimean and south Russian bases) which can now effectively limit the political and military options open to the U.S. and which would represent a real military threat in the event of conflict.

In its evaluation of what is commonly referred to as the "threat from the south", NATO takes into consideration not only the Pact's effective military capabilities, but also the eventuality that some Middle East or North African countries might allow the Soviet military forces to use their air and naval facilities or that they might even join forces with the Soviet Union in the event of an East-West crisis.

Over the last few years, parallel to the transformations in the strategic picture, not only in the Mediterranean but in the entire southwest Asia area, another important change has taken place.

The European countries of the Mediterranean no longer relate their security needs exclusively or principally to the new and heightened threat posed by the Soviet Union.

They instead tend to attribute a more explicitly national connotation to their security needs. Greater attention is being paid to eventual scenarios of conflict outside the East-West context, to which the reciprocal support clauses of the North Atlantic Treaty might not apply, and to the defence of exclusively national political and economic interests. Diffused regional instability, the deterioration of relations between many states, the growing militarisation and sharper contrasts and conflicts over the question of the use and exploitation of the sea seem to weigh more heavily than the old, traditional scenario of conflict between the two blocs.

There are also other reasons for this less "Atlantic", more "national" dimension that the Europeans now tend to attribute to their security problems - a dimension which affects not only decisions as to the structure of their military forces, (12) but foreign policy choices as well. These include maintenance of the country's capability to intervene in support of certain Third World countries (for example, France in support of the central African countries); the undertaking of new commitments as the result of international treaties (for example, Italy's commitment to guarantee Malta's neutrality); realignment of the country's international relations (for example, Turkey's shift to a policy of greater openness to the Eastern countries and the Arab world).

These developments give the impression that practically all the coastal countries are preparing themselves for a future in which, in the Mediterranean area, international tensions will be generated above all as the result of unresolved political problems and the difficulties encountered in finding solutions to the problem of "sharing" the sea and its resources, and in which the crises will be mainly south-south or north-south rather than east-west.

The increased militarisation of the Mediterranean tends to increase the area's instability; brings with it greater risks of confrontation in the event of crisis; encourages the propensity to use military instruments rather than diplomacy to resolve international controversies; and complicates control and management of crises.

In particular, the build-up of the air and naval forces of even the smallest riparian countries raises a series of problems. In the Mediterranean, the longstanding concept of high seas, to which freedom of navigation, fishing rights, exploration and exploitation of the sea's resources, etc. are closely linked, appears increasingly open to dispute.

There is a very real possibility that for alleged motives of security, navigational safety, or pollution control, limits be placed on freedom of transit in certain zones of the sea.

In the event of an open international conflict or of a domestic crisis which involves the threat of guerrilla movements, security zones might be created in which navigation would be subject to rigid control measures which might include stopping, searching and eventually confiscating ships in transit. Especially in the case of conflict between two coastal countries, large tracts of the sea might be implicitly considered or explicitly proclaimed war zones, that is, zones which are dangerous for the navigation of all types of ships, even those of non-belligerent countries. For political reasons, naval blockades or limits on transit might be imposed by neighbouring countries in support of one of the warring parties, to discourage or impede the shipment of military aid and supplies.

In addition, there might be situations in which a country decides to impede the information gathering activities of ships and planes off the shores of its coast, even beyond the limits of its territorial waters. (13)

Does this mean that the Mediterranean will eventually become a fragmented sea not only geographically but also militarily and politically? Might it lead to a different formulation not so much of the concept of international waters as of the operations which until now have always been allowed, limiting their nature and/or scope? Might it mean that the use of air and naval forces for political ends, that is, the diplomatic use of naval power, is no longer possible?

From a "technical" point of view, that is, in terms of military capability, the coastal countries are theoretically in a position, though with different degrees of efficacy, to create and control situations of the type described above.

The proliferation of high-tech weapons systems has increased the vulnerability of the surface naval forces and has caused the virtual disappearance from the Mediterranean of "low threat" areas, that is, areas in which it would be possible to operate without excessive risks and with few losses.

The Anglo-Argentine war over the Falklands in 1982 showed the effectiveness of the use of air forces in an anti-ship role and the lethality of modern missile systems. This is an element which is bound to weigh heavily in a closed and limited sea like the Mediterranean, especially if, as in the case of Argentina, the attack planes could operate from a sanctuarized territory, that is, immune to the aerial offensive of the adversary.

However, as in the case of naval forces, we must avoid the tendency to overestimate the significance of the war events in the south Atlantic. Just as it would be mistaken to affirm that the sinking of the English ships was of such great significance as to negate the validity of the use of naval forces, it would be equally erroneous to assign an absolute value to the undeniable effectiveness of the air attacks. The Falklands war proved once again that aerial forces can at times decisively affect the evolution of a conflict. But in the case in point, the price paid by the Argentine Air Force was particularly high and hardly sustainable in the prospect of a protracted war effort. The vulnerability of attack planes has grown along with that of the surface ships. It is not easy to avoid or counter the threats posed by radar-controlled, completely automatic, rapid-firing cannons and machine-guns and by infra-red or radar-guided surface-to-air and air-to-air missiles.

Thus the use of air power, too, appears more complex than the Anglo-Argentine conflict would lead us to believe if we were to consider only the outcome of counter-aviation and support missions and the losses inflicted on the British naval forces.

The fact that the coastal countries are theoretically capable of military intervention and that the entire Mediterranean has become an area of high risk do not lend credibility to the above questions nor do they lead to an unequivocally affirmative answer.

From the political point of view, it would be difficult to find sufficiently valid motives for creating or imposing situations which would limit freedom of navigation and would consequently be viewed by the other Mediterranean countries and by the superpowers as a threat to their vital interests; a threat, therefore, to which it would be impossible not to respond, especially on the part of the two superpowers, who would certainly not be willing to accept limits to the freedom of manoeuvre of their fleets.

Similar reasoning can be applied to the use of air and naval forces for diplomacy. It would in fact be rash to say that air and naval operations of the type included in what is commonly referred to as "gunboat diplomacy", or what Edward Luttwak has called "naval suasion", (14) are no longer possible.

Let us exclude from our analysis the use of naval forces for foreign policy ends by one of the superpowers to pressure the other.

Even if a country is armed with submarines, missile units and modern attack planes equipped with sophisticated weaponry, this is generally not enough to effectively deter the use of naval force by an adversary unless the country's military capabilities are supported by an adequate early warning and surveillance system and by an effective C^3 system which makes it possible, by means of a constant assessment of advantage and risk, to graduate reaction and control the eventual escalation of the confrontation. But even military power alone is not enough. It must be integrated by an equally credible political instrument capable of exploiting both the eventual weakness of the adversary and eventual international support.

But the fact that gunboat diplomacy is still possible does not mean that it has not become much more complex, risky and costly.

The political constraints to the use of military force when the risks are high could be stronger than foreseen.

The Military Presence of the Riparian Countries

The interests at stake could be perceived by the country under pressure as vital (or might objectively be vital) and therefore more important than those of the country that has chosen to use military force. But this, in turn, might act as an intrinsic dissuasion factor with respect to the latter.

The small countries might not hesitate to confront a more powerful country or use their international ties - especially their ties with one of the superpowers - as a further deterrent. This might lead to a fáilure of those conditions of sanctuarisation of air and naval space which have often facilitated the use of military use.

Situations might be created in which the naval units designed to represent the privileged instruments of "gunboat diplomacy" end up as "hostages" because of their increased vulnerability. This could increase the risks of "escalation", especially if these units are of high real and symbolic value, as in the case for aircraft carriers. A real or presumed threat to such a value could set off a reaction superior to that militarily necessary or politically desirable.

The degree to which the military instrument can be used for political ends (and hence the relative degree of difficulty and risk) depends, of course, on the overall balance of power between the two countries concerned. Apart from the two superpowers, whose real limit lies in the dangers of an involvement which might lead to a direct clash, the disparities in strength among the Mediterranean countries, in relative terms, have been greatly reduced. This means that there has been an increase in the number of countries for whom the use of naval power, either as a military or as a diplomatic tool, is no longer an easy and acceptable foreign policy option.

In conclusion, acquisition of significant military capabilities by the Third World Mediterranean countries implies a redistribution of political and military power that it would be naive to ignore or underestimate. Their "sea denial" power, which already exists and is likely to grow in the future, will have to be taken into account in any crisis which might arise as the result of controversies over the limits of exclusive economic zones and of jurisdiction over the continental shelves, over the right to freedom of navigation and transit, etc.

Of course, the existence of military potential does

224

not always mean that it is credible or applicable or that
the political conditions for its use exist.

Although the use of air and naval forces as an
instrument of coercion or pressure has become more difficult
and costly, it has still not become impossible, even though
it should be remembered that the naval forces can be used
politically only if they can survive militarily.

However, in the Mediterranean the build-up of the air
and naval forces of the riparian countries appears to carry
with it even wider-ranging implications.

The final question is in fact as previously noted: To
what extent will this phenomenon and its developent trend
affect the prospects for arms control?

The overall problem of arms control in the Mediterranean area has not yet been systematically analysed.

Ciro Zoppo, one of the few experts who has dealt with
the problem in the context of US - USSR relations, (15) has
singled out two basic categories of eventual arms control
measures. The first he calls "prudential arms control",
which includes the code of conduct currently in force concerning the two superpowers' ships and/or aircraft when in
close contact or when used in shadowing operations, the
limits to manoeuvres of naval units and military aircraft
operating near merchant ships, and all those other measures
designed to prevent a conflict generated by errors of evaluation, mistaken perceptions of threat and misunderstandings
as to the adversary's intentions. The second category,
called "substantial arms control", includes all those
measures designed to affect the two superpowers' military
presence (size and structure of their fleets) in the Mediterranean and to restrict the supply of arms to the riparian
countries.

It is evident, however, that the problem of arms
control in the Mediterranean cannot be resolved unless it is
faced globally.

If arms control measures - further tuning of the
prudential ones to improve command and control of the naval
forces in the event of crisis and effective application of
the substantial ones - were adopted by the two superpowers,
this would certainly be of great importance. It could
represent that sign of good will necessary to diminish appre-
hensions and focus the attention of other countries on the
problem. But this alone would not be enough. What is necessary is the full participation of all the European and

Mediterranean countries. Measures to limit arms supplies to the coastal countries by the superpowers in the event of a crisis could in fact have little effect if the European countries failed to adopt the same policy.

It is beyond the limits and scope of this chapter to examine what prudential and substantial arms control measures might today be adopted by the two superpowers; which of these might be effectively applicable to all the Mediterranean countries; if and in what way the measures agreed on by the U.S. and the Soviet Union could eventually be linked with regional agreements; what role the European countries might play; if it would be possible to insert explicit arms control measures in the framework of eventual negotiations on the limits of exclusive economic zones; in what way further militarisation of the Mediterranean could be limited. But these problems are too important to be ignored.

A good starting point would be the formation of a small group of "experts" with the task of analysing the arms situation, examining the prospects for arms control, and advancing a series of proposals. The results of the work of the group should be presented to all the governments of the Mediterranean area to serve as the basis for the elaboration of political hypotheses which would then be the subject of international negotiations.

The trend in all the Mediterranean countries towards a build-up of air and naval forces is sowing the seeds of greater instability and more diffused conflict. If to this is added the proliferation of nuclear arms, any further Mediterranean crisis would include the risk of a nuclear holocaust.

NOTES

1. The low level of effioiency seems to be characteristic of all three branches of the armed forces, affecting especially the more complex weapons systems: tanks,
submarines, combat planes. With regard to the status of the
naval forces, see U.S. Naval Institute Proceedings, March
1982, p. 46.

2. Cfr. Senate Delegation Report, Perspectives of
NATO's Southern Flank, April 3-13, 1980; A Report to the
Committee on Foreign Relations, United Staes Senate, June
1980, USFPO, Washington, 1980, p. 13; Marvine Howe, "Tension
over Aegean increasing", in International Herald Tribune, 18
February 1982, p. 5.

3. Cfr. Giacomo Luciani, "The international economic
importance of the Mediterranean", Lo Spettatore Internazionale, No. 1/1981, p. 16.

4. From 1 January 1946 to 31 October 1975, the United
States used its armed forces in the Mediterranean for poli-
tical ends on 63 occasions. Cfr. Barry M. Blechman and
Stephen S. Kaplan, "The political use of military power in
the Mediterranean by the United States and the Soviet
Union", Lo Spettatore Internazionale, No. 1/1978, pp. 29-66.

5) It increases from west to east from a minimum of
36.5% to a maximum of 40%.

6) On "gunboat diplomacy" cfr. Hedley Bull, "Sea power
and political influence", in Power at Sea, Adelphi Papers,
No. 122, IISS, London, 1976; James Cable, Gunboat diplomacy:
political application of limited naval force, London, 1970;
Edward Luttwak, The political uses of sea power, John
Hopkins U.P., 1974; Edward Luttwak and Robert G. Weinland,
"Sea Power in the Mediterranean: Political utility and mili-
tary constraints", in The Washington Papers, No. 61, 1979.

7) Data for this chapter have been taken mainly from
The Military Balance 1982-1983, IISS, London, 1982, integrated by other sources.

8) In 1982, the Italian air force started the con-
version of two figher-bomber squadrons from F-104 to Tornado
aircraft.

9) Another Foxtrot has been ordered and is being built
in the Leningrad shipyards while the training of Libyan
crews in the Soviet Union continues.

(10) On the average, the modern FAC have ranges which
vary from 1500 to 2000 nautical miles at cruising speed

(15-18 knots) and from 500 to 600 nautical miles at high speeds. However, Israeli FAC of the _Reshef_ class circumnavigated Africa in 1973 and crossed the Atlantic to New York harbour in 1976.

11) At the end of 1981 a contract was signed with Brooke Marine for the supply of two MLSs and an agreement has been made with Vosper Thornycroft for the construction in the Mers-el-Kebir shipyards of a 400-ton FAC.

12) France has strengthened its external intervention capacity with the creation of a second rapid deployment force, while in Italy there is talk of creating a mobile intervention force, whose tasks might eventually be limited only to the Mediterranean area.

13) Outside the Mediterranean, there was the episode of the "Pueblo" captured by the North Koreans, while in the Mediterranean, Libya's jets attacked a U.S. C-130 carrying out an electronic data gathering mission in the international waters off the Gulf of Sidra.

14) Cfr. Luttwak and Weinland, _op. cit_;, pp. 7-53.

15) Cfr. Ciro Zoppo, _Naval Arms Control in the Mediterranean_, California seminar on arms control and foreign policy, Research paper No. 57, 1975.

Table 7.1 : Spread of Anti-ship Missiles in the Mediterranean

Type of Missile	Algeria	Egypt	France	Greece	Israel	Italy	Libya	Morocco	Spain	Syria	Tunisia	Turkey	Yugoslavia
Exocet			*					*			*		
Gabriel					*								
Harpoon				*	*				*			*	
Otomat		*				*	*						
Penguin				*								*	
SS-12							*				*		
SS-N-2 Styx	*	*					*			*			*

Table 7.2 : Characteristics of Surface-to-Surface Missiles Deployed in the Mediterranean Area (Corvettes and Fast Attack Craft)

Type	Country of Origin	Range (Km)	Warhead Weight (Kg)	Propulsion	Guidance missile Warhead	Mounted on	Country
Exocet	France	42/70	165	S	Inertial ARS	La Combattante III Lazaga	Greece Morocco
Gabriel	Israel	41	150/ 180	S	BR/O SARS/TV	Saar Reshef Aliya	Israel
Harpoon	USA	110	225	S/TB	Inertial ARS	Reshef Dogan	Israel Turkey
Otomat	Italy/ France	80/100	210	S/TB	Inertial ARS	October Ramadan Sparviero Wadi M'ragh	Egypt Libya Italy

The Military Presence of the Riparian Countries

Penguin	Norway	20/40	120	S	Inertial IRS	La Combattante III Kartal	Greece Turkey
SS-N-2 Styx	USSR	42	360	S/L	AUT/RC ARS/IRS	OSA – I OSA – II Komar Nanuchka	Algeria Egypt Syria Libya Yugoslavia

Key:

Propulsion   S = solid-fuel rocket   TB = Turbojet   L = liquid-fuel rocket

Guidance   ARS = active radar seeker   SARS = semi-active radar seeker   TV = television command
IRS = infra-red seeker   RC = radio command   AUT = auto pilot
BR = beam riding   O = optical

Sources: Strategic Survey 1975, IISS, London, 1976, p. 23.
The Military Balance 1978–1979, IISS, London, 1978, pp. 96–97
integrated by other sources.

Table 7.3 : Characteristics of Air-to-Surface Missiles Deployed in the Mediterranean Area

Type	Country of Origin	Range (Km)	Warhead Weight (Kg)	Guidance Missile	Guidance Warhead	Launch Aircraft
AS-1 Kennel	USSR	100	n.a.	BR	SAHR	Tu-16 Badger
AS-4 Kitchen	USSR	450	n.a.	Inertial	n.a.	Tu-22 Blinder
AS-5 Kelt	USSR	160	n.a.	n.a.	AHR	Tu-16 Badger
AS.30	France	12	230	CG	IR	Mirage III
AS.37 Martel	France/ Britain	60	148	PHR	PF	Mirage III Jaguar
Exocet	France	50-70	165	Inertial	AHR	Super Frelon
Harpoon	USA	110	225	Inertial	AHR	F-4, A-7
Kormoran	Germany	37	160	Inertial	A/PHR	F-104, Tornado
Maverick	USA	22	59	O/TV	TV/aut. Laser	F-4, A-7

Abbreviations:

AHR = active homing radar   aut. = automatic   BR = beam-riding   CG = command guidance
IR = infra-red   O = optical   PF = proximity fuse PHR = passive homing radar
SAHR = semi-active homing radar   TV = television optical

Source: The Military Balance 1978-1979, IISS, London, 1978, p. 92, 93 integrated by other sources.

Table 7.4 : Corvettes, Fast Attack Craft, Hydrofoils Armed with Surface-to-Surface Missiles

Country	No. Corvettes	SSM	No. FAC	SSM
Algeria	2 Nanuchka	2x4 SS-N-20	3 Osa-I	3 x 4 SS-N-2
			8 Osa-II	8 x 4 SS-N-2
			6 Komar	6 x 2 SS-N-2
Egypt			4 Komar	4 x 2 SS-N-2
			8 Osa-I	8 x 4 SS-N-2
			9 October	9 x 2 OTOMAT
			6 Ramadan	6 x 4 OTOMAT
France			4 Patra	4 x 6 SS-12
			1 La Combattante	1 x 4 SS-11
Greece			4 La Combat. II	4 x 4 Exocet
			6 La Combat. III	6 x 6 Penguin
			4 La combat. III	4 x 4 Exocet
Israel	2 Aliya	2x4 Gabriel	15 Reshef	15 x 4 Harpoon
		2x2 Harpoon		15 x 5 Gabriel
			12 Saar	6 x 8 Gabriel
				6 x 6 Gabriel
			2 Flagstaff (hyd)	2 x 4 Gabriel
				2 x 2 Harpoon

Country	No.	Class	Missiles	No.	Class	Missiles
Italy				7	Sparviero (hyd)	7 x 2 OTOMAT (TESEO)
Libya	1	Nanuchka	1x4 SS-N-2	3	Susa	3 x 8 SS-12
	4	Wadi M'ragh	4x4 OTOMAT	12	Osa-II	11 x 4 SS-N-2
				10	La Combat. IIG	10 x 4 OTOMAT
Morocco				4	Kormoran	4 x 4 Exocet
Syria				6	Komar	6 x 2 SS-N-2
				6	Osa-I	6 x 4 SS-N-2
				6	Osa-II	6 x 4 SS-N-2
Tunisia				3	P-48	3 x 8 SS-12
Turkey				4	Dogan	4 x 8 Harpoon
				9	Kartal	9 x 4 Penguin-2
				4	Improved Kartal	4 x 8 Harpoon
Yugoslavia				6	Rade Koncar	6 x 2 SS-N-2
				10	Osa-I	10 x 4 SS-N-2

Forces projected to the completion of current acquisition programs

Sources: The Military Balance 1982–1983, IISS, London 1982; Jane's Fighting Ships 1981–82. Military Technology, Issue 20, 1980 and issue 23 April/May 1981. Rivista Marittima, years 1981–82, Défense National, years 1931–1982.

Table 7.5 : Characteristics of Selected Combat Aircraft
Deployed in the Mediterranean Area

Model	Country of Origin	Max Speed (Mach or mph)	Typical Combat radius (Km)
A-7D Corsair II	USA	0.87/0.92	750-825
F-4 Phantom	USA	1.2/2.27	225-1,056
F-5E Tiger II	USA	1.0/1.5	278-686
F-16	USA	1.2/2.05	550-925
F-104G	USA	1.2/2.2	1,200
G-91Y	Italy	690/0.95	370-565
Mirage IIIE	France	1.1/2.02	1,200
Mirage V	France	1.1/2.02	650-1,300
Mirage F-1	France	1.2/2.2	740-900
Mig-23/Flogger B	USSR	1.1/2.3	725-805
Su-7B Fitter A	USSR	1.2	280-400
Mig-27 Flogger D	USSR	0.95/1.6	390-805
Su-17/-20 Fitter C/D	USSR	1.05/2.17	420-600

Sources: The Military Balance 1977-78, IISS, London, 1977,
           p. 88-89
           Robert P . Berman, Soviet air power in transition,
           the Brookings Institution, 1977.
           William Green, The Observer's book of aircraft,
           London, 1981.
           Air Force Magazine, Soviet Aerospace Almanac,
           March 1982, pp. 95-102.

Table 7.6 : The Mediterranean Naval Market 1970-1981

Customer Country	Year Ordered	Building Country	Units	Type
Egypt	1977	Britain	6	Ramadan FACs (M)
Greece	1970	FRG	4	Type 200 submarines
	1970	France	4	La Combattante II FACS (M)
	1974/75	France	4+6(lic.)	La Combattante III FACS (M)
	1975	FRG	4	Type 209 submarines
	1980	Holland	2+?(lic.)	Kortenaer-class figates
Israel	1972	Britain	3	Type 206 submarines
	1979	USA	1+1(lic.)	Flagstaff Hydrofoils (M)
Libya	1976	Italy	4	Wadi-class corvettes (M)
	1975	France	2	PS-700 landing craft
	1977	France	10	La Combattante II FACs (M)
	1979	Turkey	14	SAR 33 FACs (M)
	1980	Turkey	9	C-107 landing craft

237

The Mediterranean Naval Market 1980–1981

Customer Country	Year Ordered	Building Country	Units	Type
Morocco	1973	France	2	PR72 FACs
	1975	France	3	Landing craft
	1977	Spain	1	Descubierta-class frigates
	1977	Spain	4	Lazaga FACs (M)
Spain	1972	FRG	1+5 (lic.)	38 m. FACs
	1972	FRG	1+5 (lic.)	57 m. FACs
	1974	France	4 (lic.)	Agosta-type submarines
	1977	USA	3 (lic.)	Perry-type frigates
Tunisia	1981	France	3	La Combattente III FACs (M)
Turkey	1972/80	FRG	2+3 (lic.)	Type 209 submarines
	1973	FRG	1+3 (lic.)	57 m. FACs
	1976	FRG	2+12 (lic.)	SAR 33 FACs (M)
	1979	FRG	4	38 m. FACs

In the table are not included the vessels that Algeria, Libya, Syria and Yugoslavia have received from the Soviet Union.

Source: Military Technology, 4, 1982, p. 93.

Part Four

EUROPE, THE SUPERPOWERS AND THE MEDITERRANEAN

Chapter Eight

EUROPEAN CONCEPTS FOR THE MEDITERRANEAN REGION
Elfriede Regelsberger and Wolfgang Wessels

INTRODUCTION: THE POLITICAL CHALLENGES

As basic pillars of the post-war system are being eroded, West European states - and especially the ten members of the European Community - are becoming more and more aware of the growing international challenges they face and have consequently resolved to "seek increasingly to shape events" in the international system. (1) Since the early seventies the European Community and its member countries - hereafter referred to as the EC - have proclaimed on various occasions and in different forms that "Europe" has to establish a "European identity" in world affairs. (2) In these declarations the EC has elaborated carefully drafted principles to indicate the European interests and political values it intends to pursue and uphold. Political parties, members of the European Parliament, social forces and academic circles have contributed to the debate, which is, however, far from leading to a West European foreign policy consensus or even some form of clear policy alternatives. Discussion of a possible new international role for the EC is quite often fragmented and inconsistent, problems are artificially separated from one another, and priority is attributed to just too many areas and questions, without developing the means and instruments necessary to achieve the stated goals.

The highest-ranking officials of the EC have attributed "essential importance......to the fulfilment of its commitments to the countries of the Mediterranean Basin". (3) The EC Commission has recently stressed that the "Mediterranean is an area of considerable importance to which the Community, for historical and geographical reasons, is in a unique position and has responsibilities it cannot shirk". (4) To secure peace in Europe as well as to stabilise trade flows, the Community has to contribute to the solution of conflicts in northern Africa, the eastern Mediterranean and the Middle East. (5) "Europe must play its rightful role on the international scene and particularly in the Mediterranean. To assert its presence and strengthen its identity, the Community must equip itself with the instruments of diplomacy and joint action which it still lacks at the level of policy and security". (6)

Behind these general statements, different perspectives and motivations can be discerned. The Christian Democrats in the European Parliament (EP) have stressed the importance of a Mediterranean policy for the purpose of containing Soviet expansion while the Gaullists want to use the EC to exclude both the USSR and the USA from this area. They point to the common emotional, cultural and historical links in this core area which had given birth to European civilisation. The Socialists in the EP have demanded that the process of détente be extended to the Mediterranean region as well. (7)

THE HISTORICAL DEVELOPMENT OF AN EC MEDITERRANEAN POLICY: A DIFFERENTIATION OF INSTRUMENTS BUT NO MAJOR CHANGE IN THE ECONOMIC STATUS QUO

In the relationship between the EC and the Mediterranean countries certain periods can be distinguished which show an increase in activities and a differentiation of instruments and policies. The new agreements have served, however, mainly to maintain basic elements of the economic and political status quo. No new division of labour has been established.

The first phase of Community policies
The Mediterranean area was one of the first regions to which the "young" Community turned in the early sixties.

With the Association agreements concluded with Athens in 1961 and with Ankara in 1963 (8) the Community developed its first formal links with third countries. These steps were motivated by general political interests (U.S. pressure included) to prevent any "decoupling" of the southern European NATO members and, after the breakdown of the first membership negotiations with the United Kingdom, to soften the impression that the EC was a "closed shop" of certain West European countries. The concept of the two Association agreements was the eventual integration of both countries into the Community. It was thus assumed that a rapid convergence of economic performances and structures could be achieved. However, as the Association agreements evolved, especially in the case of Turkey, this basic notion brought significant and increasing difficulties for both sides.

The second phase of EC Mediterranean policy involved the conclusion of nine further bilateral agreements which were signed in the second half of the sixties. The treaties with Israel, Lebanon, Morocco, Tunisia, Spain, Malta and Cyprus as well as those with Egypt and Portugal were restricted mainly to the establishment of preferential trade relations.

None of these agreements was geared to a membership position; some of them, especially the more important ones, served to lead from colonial times to a post-colonial era keeping intact the basic characteristics of the established trade patterns. (9)

In the third period, the early seventies, the Community developed the concept of a "global Mediterranean policy" guaranteeing free access for industrial products to the European common market and facilitating agricultural exports from the Mediterranean basin to the Community market, offering financial aid and envisaging cooperation in a broad range of areas, including environment, fisheries, etc. By 1978 a complex network of agreements between the European Community and nearly all Mediterranean littorals (with the exception of Libya (10) and Albania) had been established.

In this period, the Community itself was undergoing changes. With the 1969 summit in the Hague and especially the 1972 summit in Paris the Community had developed a new action programme which included the goal of strengthening its international role: "Europe must be able to make its voice heard in world affairs, and to make an original contribution commensurate with its human, intellectual and material resources". (11) This "original contribution was soon perceived as being based on the concept of Europe as a "civilian power". (12)

According to this concept, the Community was to make

full use of its economic potential by establishing equal
partnerships with third countries or, even better, with
groups of third countries. By foregoing superpower attributes like military force and by going beyond traditional
colonial relationships, it was believed that the Community
would be able to become a leading force in an evolutionary
and guided change of the international system based on
democratic and human values. A global Mediterranean policy
was conceived to be, along with the Lomé agreement, a central
part of this role as a civilian power.

European Political Cooperation and the Mediterranean Region

Not only this third phase of Community policies (in a
strict sense), but also a new West European mechanism for
coordinating foreign policies had an impact on the Mediterranean area.

In the early seventies the former six and now ten EC
member states developed a new form of involvement in the
international system: the intergovernmental cooperation
procedure among the foreign offices of the member countries,
called European Political Cooperation (EPC) (13) (which is
distinct from the external relations of the EC provided for
in the Rome Treaties). The objectives of this mechanism for
consultations on foreign policy issues are as formulated in
the Luxembourg Report (14): "to ensure, through regular
exchanges of information and consultations, a better mutual
understanding on the major international problems; to strengthen their (that is the member states') solidarity by
promoting the harmonisation of their views, the coordination
of their positions, and, where possible and desirable,
common actions".

From the outset, the Mediterranean and Middle East
region was a major focus of EPC. Along with the foreign
ministers' declarations on the Greek-Turkish conflict over
Cyprus in the mid-seventies, the consultations on the political developments in Portugal after the revolution, and the
statements of the Nine concerning Spain under Franco's dicta-
torship and afterwards, the Middle East conflict was a
central topic at various meetings at various EPC levels. On
several occasions the foreign ministers of the EPC adopted
common positions on the most delicate points of the Middle
East conflict, striving for a balanced approach between Arab

and Israeli interests. Over the years consultations within the EPC framework have covered the situation in Lebanon and the Camp David process. In 1980 the European Council launched a declaration on the Middle East (Venice Declaration, 13 June 1980), assuring the Nine's readiness to play an active role to establish and guarantee extensive and lasting peace in the region and stressing the need for the PLO's participation in future negotiations. As in former times, the European diplomacy towards Israel and the Arab states met with strong reservations from Jerusalem and, more gravely, from Washington. (15)

To complete the list of EPC activities in this region, mention should also be made of the Nine/Ten's declarations concerning the Iran-Iraq war since 1980 and the roles of Syria and Jordan in it, and the discussions at the ministerial level on developments in Turkey after the General's coup. (16)

The Euro-Arab Dialogue and Concepts for Group to Group Negotiations

Besides the global Mediterranean approach of the EC based on (bilateral) economic measures, the EC member states combined elements of the EPC and EC machinery to start a new form of interregional cooperation between the European Community and the member states of the Arab League - the Euro-Arab dialogue (EAd). After the United States had made it clear in 1974 that an active European role in the Arab-Israeli conflict might undermine Kissinger's Middle East policy, the Nine decided to establish a long-term partnership with the Arab world leaving these questions officially aside. In order not to undermine the American plan of a coordinated Western energy policy, which led to the creation of the International Energy Agency in 1979, the EC also excluded energy issues from the dialogue.

In practice, however, the exclusion of political topics proved to be artificial. Since the inaugural session in Paris in 1974 the representatives of the Arab League have stressed the need to discuss recognition of the Palestinians and Israel's policy since 1967. (17) The EAd has had its ups and downs, almost collapsing at a certain point. Nonetheless, in 1980 the nine heads of state and government stressed the importance of the EAd "at all levels" especially the development of its "political dimensions". Since then several meetings of the so-called "ad hoc group" (made up of representatives of the Arab League and of the EC) have taken place to organise a conference at the level of foreign ministers. A final date has not yet been fixed because of the differences of opinion concerning the PLO. (18)

The Euro-Arab dialogue constitutes a special form of diplomacy: it brings together two organisations of states, the European Community and the Arab League. The dialogue has shown signs of progress in areas relating to industry, infrastructure, and agriculture and in cultural-scientific matters. Controversial positions still exist, however, over the opening of the EC market to Arab products, the question of capital investment, and the status of Arab guest workers. (19)

The more the energy questions became important in the seventies the more the proposed concepts for a "Mediterranean" policy came to include countries like Saudi Arabia and Iran. (20) Among the points of discussion were a "trialogue" between the EC, the Arab and the African countries as proposed by the former French President Giscard d'Estaing and Bonn's idea of cooperation agreements between the EC and the countries of the Persian Gulf region including Iran, Kuwait, Bahrain, Qatar, the United Arab Emirates, Oman, Saudi Arabia, and North Yemen.(21)

Towards a New Phase?

In the second half of the seventies developments on the northern shore of the Mediterranean changed the political map: the democratisation of Greece, Portugal and Spain opened the way for their membership in the Community. The southern enlargement process was set in motion and is creating a new framework for the EC's Mediterranean policy. The expected dynamic consequences of Portuguese and Spanish membership threaten to reduce the access of traditional Mediterranean exports to the Community. The cumulation of economic effects and internal and external political repercussions is causing considerable concern in the southern Mediterranean countries that will be affected and in the EC Commission. In June 1979 the EC Commission presented a report to the Council of Ministers which pointed to the consequences of an enlarged Community on relations with non-member countries. (22) On 24 June 1982 another document was adopted stressing the need for reviving and strengthening the overall Mediterranean approach on new bases. The motivations are linked not only to the accession of Spain and Portugal but also to the current state of EC policy towards the Mediterranean littoral countries which "is not operating to the satisfaction of our partners", as the Commission's communication states. (23)

The Commission also reformulated the principles underlying the EC's development policy, (24) putting the Mediterranean policy into the broader framework of the EC's relations with the "South". The Mediterranean policy is always mentioned <u>after</u> the Lomé convention. The principles of this approach stress assistance aimed at fostering autonomous growth and development of the rural economy in these countries as well as the need to work together with the developing countries in formulating and implementing this policy.

At the same time, the assumptions of the "civilian power" concept on which the Mediterranean policy was based have been called into question by various developments. The economic power of the EC has clearly declined as a consequence of higher energy prices and persistent economic crises. Consequently, the EC's Mediterranean policy, devised in the economic boom years of the 1960s, can hardly be applied properly in now different circumstances. The room for manoeuvre has been drastically reduced, and instead of opening its economy further, the EC has begun to adopt protectionist measures. These "adaptations" of the EC policies have created major problems for its partners, who had based their economic plans on different expectations.

The Soviet invasion of Afghanistan has undermined the détente process, which was another implicit assumption of the civilian power concept. Military and security problems, rather than lose their importance as it was hoped during the heydays of détente, re-emerged with greater force in the early eighties. Disputes over regional balances in the southern parts of the Mediterranean region likewise increased (Libya's opposition to Egypt's policy, the tensions between Morocco and Algeria over the Western Sahara, the Iran-Iraq war). Both global and regional developments are increasingly challenging European policies which have tried to avoid direct involvement in military and security questions. (25)

European Concepts for the Mediterranean Region

THE IMPORTANCE OF THE MEDITERRANEAN REGION FOR THE EC

The importance of the Mediterranean for Europe is usually deduced from the member countries' interests in six major problem areas (26): a) the commercial aspect, especially in regard to export markets for the Community; b) the security of energy supplies for Western Europe; c) the security aspect, especially the open southern flank of NATO and possible spill-overs from the Middle East conflicts; d) political stability in the Mediterranean in relation to the maintenance of the status quo and the preservation of peace; e) the cultural-historical heritage; f) the management of common problems in the Mediterranean area.

How vital are these interests and what role can the EC play in these questions?

Leaving aside the commercial aspects, which are well known, and the problem of energy supplies, which is discussed in other chapters, we shall focus in this section on the other aspects just mentioned.

General Political and Security Interests

Political stability in the Mediterranean and a guarantee of military security in the region are perceived as major aims of EC and Atlantic Alliance policies. (27) This aspect has always been one of the motivations for the EC Mediterranean policies from the Association treaties with Greece and Turkey in the early sixties - which reinforced their integration into the Western coalition - up to the agreement with Yugoslavia in 1980 which gives West European support to an independent Yugoslavia in the post-Afghanistan and post-Tito era. With the creation of EPC, the Nine/Ten have tried to influence political developments in Cyprus, the Middle East, Spain, Portugal and Turkey. The Euro-Arab dialogue had similar implications, but the direct links of EC and EPC instruments to the stability and security of the Mediterranean region are, as pointed out above, limited. In the superpower constellation a civilian power is of only limited value if the conflicts become military. The rather marginal role the Ten have played in the Middle East since their Venice initiative is one sign of their limited influence. As the struggles for regional dominance increase and intensify, the European policies toward them might also be of limited value because of the "balanced" approach adopted. Bilateral policies (e.g. Italy's security guarantee for Malta, France's and Italy's involvement in Lebanon) might ultimately prove more important.

Greece's entry into the European Community, and hence into EPC, on 1 January 1981, posed new problems and gave proof of the scarce effectiveness of European Political Cooperation in the Mediterranean area. Full Greek membership automatically entailed a greater "involvement" of its partners in the Greek-Turkish disputes (over the Aegean Sea as well as Cyprus). Aware of this limited capacity to deal with military and security matters - which are central to the Greek-Turkish conflict in the Aegean (as exemplified by Greek demands for guarantees of the security of its borders with Turkey) - the Nine chose to take sides with neither Athens nor Ankara, offering to play ony an advisory role, if asked. After Papandreou was elected to head the Greek government, it became increasingly difficult to forge common positions on various EPC topics; for instance, on the Middle East conflict (the participation of four European countries in the Sinai peacekeeping force, voting in the UN General Assembly). Although as a result of Greece's full membership in the Community the EPC Mediterranean working group has not met since November, the EC states are well aware of the importance of the conflict and keep abreast of developments in the Greek-Turkish relationship (through meetings of EC foreign ministers, informal talks between European embassies in the Mediterranean, NATO summits). There is a strong desire among European states to stabilise Europe's position in the eastern Mediterranean, especially after Afghanistan and the Syrian revolution. Political and economic support has been given to Turkey to strengthen its ties with Western Europe in various fora (OECD summit in Guadeloupe, the EC, the Association Council, Turkey-EPC consultations at both ministerial and senior official levels). Although Turkey's full membership of the Community, as foreseen in the Association Agreement, is presently out of the question (especially due to the state of the economy and of democracy in Turkey), alternative forms of linking that important NATO member closer to Europe need to be found. A new partnership in the EPC framework could be a practicable solution to stress the importance the Community generally devotes to that region. (28)

The general reluctance of the Ten to discuss military

and security matters at all within their Political Cooperation framework also reduces the EC's actual and potential
international role. In their latest document (the London
Report) - describing the working procedures of the Ten in
the EPC framework - it was agreed that the "political
aspects of security questions" would be included in the
Ten's consultations. In the current debate over the way to a
European Union, the security issues (not to mention military
questions in the proper sense) are quite controversial and -
as it looks now - will not lead to any increased coordinated activities in this field.

As for the future, destabilising effects of possible
protectionist policies of the Comunity after enlargement can
endanger the political and social stability of the Mediterranean region thus opening the way for stronger influence of
other countries, especially the Soviet Union. (30) Moreover,
certain countries of this region will raise increasingly in
importance from strategic points of view. Turkey, Syria,
Israel, Egypt, and perhaps Algeria, belong to this group to
which the EC needs to turn special attention. Undoubtedly,
Turkey will press upon her EC - and Nato - partners to
become either a member of the Community or get special
treatment for her services within the Atlantic Alliance or
even - beyond Nato - in the whole Middle East region. (31)
Though the Community and the EPC are reluctant to deal
directly with defence matters, the Europeans will not be
able to keep themselves out of the overall security questions and the role it will be asked to play directly or
indirectly.

Cultural-Historical Links

One major factor always mentioned in connection with
Mediterranean policies is the common cultural heritage of
all Mediterranean countries stemming partly from the colonial past, partly as the consequence of an ancient tradition
based on common ideals. As Claude Cheysson has pointed out,
"from Alexander the Great to the Euro-Arab dialogue we have
both experienced so much of one and the same life". This
common heritage is difficult to define as there are quite a
lot of divergent and convergent elements; looking at the
differences in religion and in political, administrative and
economic cultures there seem to be some divergencies, though
at the same time the existing cultural and personal links
should not be underestimated. It would seem appropriate to

cultivate and develop these common links. However, the EC policies in this respect are practically non-existent. A few conferences on the subject in the framework of the Euro-Arab dialogue, though certainly useful, are not yet adequate instruments to promote these cultural links.

The Management of Common Mediterranean Problems

In the Mediterranean Sea certain problems exist which call for common management or at least a set of common rules. Topics like the environmental quality of the Mediterranean waters, transport regulations, overfishing and mining of seabed minerals are becoming increasingly important. The EC as a whole has so far shown less interest in contributing to the solution of these questions on its southern boundaries, though some areas were mentioned in association treaties with Mediterranean countries. Neither the accession of Greece nor the second enlargement of the EC seem to have had repercussions on the list of priorities in Community policies in Brussels. Controversies among the Nine/Ten about EC competences as far as pollution or the delimitation of water boundaries is concerned and their inability to find common positions (i.e. UN Conference on the Law of the Sea) are obstacles for developing coherent concepts as well as for co-operation between the EC members on the northern shore and third countries on the southern shore of the Mediterranean Basin. In the framework of the Barcelona Convention of 1976 the participation of several EC member states seems to be of greater importance and efficiency for this kind of regional co-operation in the Mediterranean than Community policy. (32) The EC as such has so far developed few policies directly oriented towards these Mediterranean problems.

The Community's environment and fishery policies will of course have a certain impact on the Mediterranean; it is doubtful, however, that these rules developed for conditions different from those of the Mediterranean region can be adequately applied. New policies and forms of co-operation will have to be worked out. (33) So far, interregional fora (i.e. the Euro-Arab dialogue) have not been used for this purpose.

Some Conclusions

On the basis of an analysis of the EC's interests in

the above areas, in the commercial field, and in the security of energy supplies to Western Europe, it could be concluded that it would seem best for the Community to be oriented toward preserving the status quo without increasing its involvement too much. It is a situation in which the EC might be less inclined to pursue constructive evolutions than to maintain the status quo at low costs.

This "realistic" view which takes into account only immediate and concrete interests might, however, prove to be shortsighted. Mediterranean policy should be looked at as an attempt to increase the Community's scope of action in this area and beyond. (34) It is a "potential" instrument, the limits of which have not yet been tested. An increase in the scope of West European action here is relevant to the two dominant conflict patterns within the international system: the Mediterranean is in fact one of those crucial areas in which East-West and North-South conflicts overlap. At least as long as the EC is interested in remaining a power capable of "shaping" the international system, it needs to conceive and implement a common policy toward the Mediterranean. If the EC is not capable of making its influence felt in its own "backyard" then the claim to pursue "worldwide" interests and responsibilities is rather badly founded. (35)

Another argument for giving the Mediterranean countries a high priority is the possibility that they look for alternatives. The validity and implications of this thesis need to be analysed. A reorientation of some trade patterns, for example, towards the Middle East markets, would not necessarily be negative. Turkish exports to Saudi Arabia could have bridge-building effects and could also increase the welfare in Turkey with positive effects on its capacity to import goods from the Community. A commercial reorientation of the Mediterranean countries toward Eastern Europe, which is considered to have negative political effects, seems unrealistic for the time being given the adverse economic situation there. What seems of higher importance to the EC than the commercial aspect is an eventual political reorientation of the Mediterranean countries. The southern and eastern Mediterranean countries have considerable political and security "nuisance value": they may not necessarily be "strong" in comparison with other powers, but they can create quite a lot of political and security problems for the EC, as countries like Libya have demonstrated.

The EC Internal Dimension

So far we have been looking at European interests as if from the 13th floor of the Berlaymont (Commission headquarters) in Brussels, i.e. assuming some form of common interest within the unit known as the European Community. However, in view of the political decision-making process which produces European policy and attitudes toward the southern Mediterranean countries, it would perhaps be more appropriate to look at the national interests of the member countries and analyse what kind of package deals are possible. A look at the trade patterns will reveal that in terms of exports the southern Mediterranean region is of relatively low importance for Denmark (about 2.5%), the Netherlands (about 4.0%), the United Kingdom (about 4.2%) and the Federal Republic of Germany (about 4.7%). The southern members of the EC are instead confronted with an economic dilemma: the southern Mediterranean countries are a profitable export market but they are also competitors in certain Community markets in which no growth can be expected (southern agricultural products, textiles). For all EC concessions to the southern Mediterranean countries, the southern members of the EC will present the bill to the Community and thus also, and especially, to the richer northern member states. These countries might find themselves confronted with a threefold demand: they will be asked to "pay" first in the form of an increased regional fund or of a new Marshall-type plan, then for a new deal in the southern and eastern Mediterranean countries, and then for compensating the southern EC members for the costs incurred as a result of enlargement and a new Mediterranean policy.

GLOBAL OPTIONS

First Set of Options: A "Mare Nostrum" of the Littorals

Looking at these constellations and interests, there is no "natural" solution to a new Mediterranean policy which would be the logical and inherent continuation of present policies and the implementation of unequivocal and dominant European interests. We have, therefore, to confront ourselves with overall options, their advantages and disadvantages, and strategies to implement these options.

European Concepts for the Mediterranean Region

In the following first set of options we assume that the EC will give priority to the Mediterranean area in its foreign policy, reducing the intensity of its relations with other areas of the world. On the basis of an analysis of its interests, the EC will seek to play a direct and active role in finding solutions to the problems in that region, proposing long-term solutions with a high degree of institutional and economic commitment on its part. In this set of options, the EC of 12 would decide to build a new regional bloc around the Mediterranean in order to tackle the existing problems. As proposed by a good number of political forces in countries around the Mediterranean, all powers "foreign" to this region - the USA and the USSR - should reduce their influence to the benefit of all "original" Mediterranean littorals. This kind of option can be pursued in the security, diplomatic and economic fields which are mutually dependent. The exact form can be differently shaped. This setting of priorities for the Mediterranean area by the EC implies - as we shall see - an analogous reorientation of the southern and eastern Mediterranean countries.

Option 1: a regional system of collective security. In this option it is assumed that the influence of the superpowers in the field of military and overall political questions would be reduced. Only the Mediterranean countries as a group would decide about the security problems in "their" regions. It is to be hoped that with the exclusion of superpower rivalry, the level of conflict would decrease dramatically. The remaining problems could then be solved by a system of collective security, in which all countries would react commonly against aggressors and in which solutions would be found by continuous negotiating among equal partners. Analogous procedures as put down in the Charter of the League of Nations and partly in the Charter of the United Nations would be established. A Conference for Security and Co-operation in the Mediterranean (similar to the CSCE, but excluding the superpowers) (37) would be a first step in this direction.

For the EC two possible ways would be open: to participate either as a unit or separately as nation states excluding the northern European countries. In the first case, the EC would need to acquire more powers and competences in the security field and get deeply engaged in the Mediterranean area. This could mean "Mourir pour Beyrouth" for a citizen from Copenhagen. In the second case, the Community would be split into a Mediterranean group and a non-Mediterranean one in crucial questions. Although some form of co-operation and division of labour could be envisaged by which the northern countries would support the Mediterranean ones (e.g. through trade sanctions, arms embargoes, etc.), the different outlooks would hinder the European countries from becoming a coherent political actor in the international system. The "wall" would not be built through the Mediterranean region but through the Community and perhaps even through some member countries like France and Italy.

This regional system of collective security is neither a realistic nor a desirable solution. The strong conflicts are not only induced by the superpowers but are endogeneous, between adjacent countries. The level of conflict might even increase, if superpower protection were reduced. (38) Furthermore, many conflicts in third world regions like the Middle East or subsaharan Africa cannot be disassociated from the Mediterranean countries so that there is no natural closed Mediterranean security area.

Even more, the process to reach such a system would lead to conflicts with the superpowers and would increase the level of tension. For NATO members questions of alliance would be raised, e.g. "Are the NATO obligations of Italy compatible with the 'logic' of a Mediterranean collective security system"? Other non-Mediterranean interests of the EC might thus be directly hurt.

If we look to history, we will find that a regional system of collective security as a means to settle conflict does not hold much promise as the assumptions necessary to make this model workable are seldom met and would certainly not be fulfilled in the case of the Mediterranean region, where there exists neither the political will nor the military capacity of the Mediterranean countries to fight collectively against regional "aggressors", whoever they may be. Coalition building in the Mediterranean could replace the collective security system. If the EC participated as a unit, the necessary balance of the model would be destroyed to the benefit of the northern shore of the Mediterranean.

Option 2: an Arab-European duopoly. This option would be based not on a collective security system, but on conflict management between two groups - the EC on the one side and the Arab League on the other. The Euro-Arab dialogue could be the nucleus of this duopoly. Countries outside the two blocs (like Yugoslavia, Turkey) would need to settle for one side or be left outside. As internal prerequisites for this option, both sides would need to become more coherent units with a higher degree of common interests and centralization than has been established so far.

Though this kind of oligopoly looks more possible, it is again neither a realistic nor a desirable solution. Apart from the remarks made for Option 1 about the exclusion of the superpowers, it seems uncertain if this duopoly really could achieve a zone of freedom and prosperity taking into account the different interests of the countries within, and on both sides of the Mediterranean coast. Furthermore, the building of blocs might - following group dynamics - increase the tensions between the groups in the middle of the Mediterranean.

Option 3: a common Mediterranean market. Leaving the security issue aside, another option for the EC lies in the building of a close economic bloc around the Mediterranean. The EC is internally better prepared for this kind of option than for security questions. The economic bloc building could take various forms. (39)

The model of the Common Market, including a customs union with a common external tariff, free movement of production factors, trade agreements with other groups, monetary co-ordination mechanism, etc., could be transplanted to the Mediterranean area.

For this option, the Community would have to either include the countries for the southern and eastern Mediterranean coast into its membership, thus forgetting its European vocation according to Article 237 of the Rome treaty, or split up letting its southern members participate in a different group, i.e. that of the Mediterranean market. Though forms of harmonising both circles could be imagined, this option would mean a clear division of the Community.

Besides this point, a Common market option again is neither realistic nor desirable. The prerequisites for a Common Mediterranean Market are not fulfilled, as the level of economic development and the conceptions about the "economic order" (capitalistic or socialist) are too divergent (40) to enable a balanced membership. The political requirements - especially to give up certain competences and powers to common institutions - will also not be met as a sufficient level of mutual confidence, and the willingness to give up sovereignty do not exist. Furthermore, the potential of a Common Mediterranean Market is not promising, at least not for the European members. And it is doubtful that a Common Market for only certain products (fruit and vegetables) (41) would be more workable.

Option 4: a new division of labour. Critics of the present asymmetrical relationship (42) propose a new division of labour by which: a) the developing countries of the Mediterranean region would diversify their one-sided export structure by investing in a broad range of labour intensive semi-processed goods, reducing at the same time their dependence on agricultural imports by intensifying their own farming of basic foods. The developing countries should also seek to become competitive and hence more self-reliant and autonomous, in more sectors. To achieve these goals, intra-regional co-operation within the southern Mediterranean and Arab world needs to be intensified; b) the Community would open its markets to the sensitive products of the Mediterranean countries, concentrating their production scope on high technology goods for which the Community industries have comparative advantages.

It is hoped that the result of this new economic division of labour would be a decrease in the asymmetrical relationship to a different form of interdependence based on a new kind of complementarity of production patterns. For this option a global convention between the Community and all Mediterranean countries analogous to the Lomé Convention as proposed by the Commission (43) could be envisaged.

Although the authors of these proposals are following a general trend of conceptualizing the new international economic order, they do not indicate that this new division of labour would probably imply new protectionist measures in and for this area. It seems to be very likely that the rather weak competitive power of the southern Mediterranean countries with regard to dynamic growth centres elsewhere in the world and the limited absorption potential of the EC make some common external protection of the whole region necessary, if this new division of labour should be established within the Mediterranean area. Otherwise, the opening of the EC market will be to the benefit of other third countries. In an open system, the goal of a Mediterranean policy as stated above could probably not be achieved.

Furthermore, the prerequisites on both sides of the Mediterranean need to be analysed:

On the side of an enlarged Community, certain industrial branches (especially textiles, food processing) would need to be replaced by new growth industries. As these branches are concentrated in already relatively backward regions of the EC and especially in the new southern members, the economic and thus political costs will be quite high within the Community; in this option it is thus necessary to restructure exactly those industries of the less-developed EC member countries in which they still have some comparative advantages within the EC. The loss of these advantages would not be acceptable to these member states nor to those political forces in the Community who want to protect the declining industries.

Internal structural changes would also have to occur in the agricultural field. In order to overcome the negative effects, the richer countries and regions would have to offer at least substantial compensation for the producers of the out-phased goods within the Community, e.g. in the form of a new Marshall fund.

At the same time, the Community would also have to raise money for investment programmes in the southern Mediterranean region, (because otherwise few chances for an economic success in the southern riparian countries will be created). These two financial demands are likely to overload the Community budget and to overstretch the willingness of the northern European countries to pay. Thus, the internal EC constellation is not favourable for such an option.

For the countries of the southern Mediterranean, this option also calls for a drastic change in the existing production and trade patterns, and it might finally mean the willingness to be integrated into some form of common European Mediterranean investment planning. It is doubtful whether these countries will be prepared to co-operate closely and to face this kind of dynamic process without relying on some strong form of own control and safeguard measures.

Option 5: Afro-Arab-Euro trialogue. Given the close economic interdependence, the geographic proximity, the common cultural values, and the complementarity of energy resources, technology, and human as well as material resources, the proponents of a "trialogue" suggest enlarging the area of co-operation to a trialogue between Europe, the Arab, and the African countries. (45) At the most basic level, this co-operation should bring together Arab capital, European know-how, and African labour forces. This trialogue could eventually serve as a model for a North-South dialogue - as assessed by some - or a closed trading bloc - as others fear. Beyond this general notion of complementarity, the forms of co-operation on a macro-economic and macro-political level are rather vague. Though the EC has formal links with the African and Mediterranean countries, it lacks ties with some of the most important Arab countries which are at present not included in the framework of Community treaties and which quite often show only limited interest in such co-operation.

To achieve more than ad-hoc co-operation on certain concrete projects, which can be and is done without any large framework, it is necessary to establish more intensive co-operation leading to a working institutional set-up and global political package deals. (46) Certain economic independences will not be enough to overcome quite large differences in economic development, political ideologies, conflicts of interests, and pressures from outside. Many of the problems mentioned in the other options will be increased considerably. Only the financial resources of some Arab countries seem to make the difference. It is, however, rather unlikely that these rich countries would invest economically and politically in the region of the trialogue to make this complex organisation work.

Second Set of Options: Diversifying EC Policies

The options we have covered so far we have assumed that the European would give priority to the Mediterranean region. Four major elements, conditions or consequences were common to these options: a) The options were "global", i.e. they tried to establish common frameworks or rules for a broad area of problems; b) the southern Mediterranean countries had to agree and to participate in the solution;

c) this set of solutions would have to overcome strong
international obstacles; d) the EC would have to either
"mediterraneanize" or be split.

In the upcoming options, we will deal with a diversification of Mediterranean policies. Whereas in the first set
of options certain overall or global approaches were suggested, the EC could also pursue forms of diversifying its
policies. Two dimensions have to be considered: the search
for solutions to problems using a pragmatic, piecemeal,
functional approach and treating countries or groups of
countries differently.

Option 6: functionalist Mediterranean policy. In this
option, the EC countries would pursue a pragmatic piecemeal
policy which would imply flexible procedures. The existing
problems, which are assumed to be disparate and not mutually
connected, would not be solved in a global and common
framework with certain legal qualities but on an ad-hoc
basis adequate to the characteristics of a given problem.
Environmental problems would be dealt with in different
groupings and in other political and legal settings than the
social questions of migrant workers. The form of management
would follow the function. (47)

For quite a few problems, this option would not
require direct Community involvement, as this might imply
unnecessary stiffness due to the legal character of Community policy. Forms of "multi-tier solution" might be envisaged. These concepts of a multi-tier Community (abgestufte,
differenzierte Integration, L'Europe à deux vitesses,
L'Europe à géometrie variable) have been intensively discussed in the last years without yet offering a satisfactory
solution to the dilemma between the necessity to find
differentiated methods of problem solving in the large
Community of twelve, which is due to objective differences,
and the equally important need to keep a sufficient coherence among member countries and to preserve the Community
character, on the other side. According to this option,
certain Mediterranean problems like the environment or fisheries might not necessarily follow all the Community rules on
environment or on fisheries, but exceptions would be decided
by the respective Community bodies.

In those fields where Community policies exist and
where unique approaches are necessary, like in the trade or

the agricultural sector, the Community needs to be involved in any case.

This option has the advantage for the Community that it may wait and see if, and in which fields, problems become politically sensitive, and then it can play the role of a fireman. The EC does not need to elaborate new structures and commit itself to an uncertain future.

The disadvantages on the other side of the coin are that conflicts, when on the table, might be too large and politically too sensitive to be solved using a functionalist approach. Many problems are interdependent or are put politically into some form of package deal. As shown by experience in other fields like the Atlantic relationship or in dealing with the USSR, the functionalist approaches of the EC are too narrow to deal adequately with overall demands embracing different policy sectors like security, diplomacy, and economy. In the Euro-Arab dialogue or in the Conference on International Economic Cooperation in Paris, the Arab countries have already confronted their European partners with this kind of demand for broad packages to which the EC found it difficult to respond adequately.

Furthermore, the EC would have no stable and reliable framework to deal with these problems. This deficit is especially negative for a political system like the Community, which has no efficient and rapidly acting decision-making centre, but which needs time-consuming internal negotiations.

Option 7: the principal nations' approach: diversifying according to partner countries. In this option, the EC would differentiate policies according to the Mediterranean littoral countries involved. The basic assumption is that the economic structure and performance as well as the political relations between the EC and the Mediterranean countries are not so homogeneous as the present global approach doctrine indicates, but different to a degree which makes individual treatment imperative. There can, therefore, be no trend towards the building of a coherent group among eastern and southern Mediterranean countries. The Community should deal with each country or small grouping separately and without a common framework. As the agreements with southern Mediterranean countries would be competitive with each other, involving hierarchies or preferences, (48) the

Community would be increasingly obliged to set priorities among its "clients". The Community might thus look for "key-states" with which a broad intensive co-operation could be reached, whereas with other - less important - countries, the agreement might be limited to certain narrow problems giving these countries a much smaller niche in the Community market.

If economic and military security dictate European interests in the Mediterranean region, the NATO member Turkey as one of the important countries should attract the Twelve's attention, and broad arrangements including perhaps even participation in EPC should be pursued, whereas the involvement of Political Cooperation would be excluded for other non-EC Mediterranean countries. (49)

With this "principal nations option", the Community would more and more interfere in struggles for regional predominance, favouring certain countries to the detriment of others. If there were strong commitments to certain countries, the propensity to support certain regimes or political forces would increase. This option would, therefore, be a highly politicized issue and might lead to a direct Community interference in regional quarrels and to competition with the superpowers. In order to reduce to a certain extent this kind of conflict potential, the Community countries could arrange a "division of labour" by which individual EC members would support, by national means (bilateral development aid, arms exports), different countries. The coherence of an EC policy might, however, be lost, and the efforts of individual countries might counterbalance each other.

A need for decisions?

The present status quo of Mediterranean policies is, at least in the commercial field, not unfavourable to the Community. In view of the other more "pressing" problems of reforming the CAP and redistributing the budgetary effects, of keeping the international market open and balancing trade with Japan, etc., many politicians might doubt that the EC can now afford the "luxury" of reviewing the present agreements and discussing new options and approaches. These overburdened politicians might consider these questions an artificial stirring up of new quarrels which it would be better to leave off the agenda for a couple of years.

European Concepts for the Mediterranean Region

Finally, they might argue that the membership of Spain is still some years away and that perhaps in the late eighties a global economic boom or economic miracles in some Mediterranean countries might ease a lot of the pressure. "Leave this question to our successors", they might argue, "now we have to tackle the problems of the mandate and of enlargement; our relations with the U.S. and the East-West conflict take precedence on our agenda; if we always have to take the Mediterranean problems into account, we will never solve any of our problems".

This line of reasoning holds some truth inasmuch as it underlines the fact that not all problems can be resolved at the same time and that a hierarchy of problems needs to be established. At the same time the awareness of these − perhaps − secondary but certainly persistent and crucial problems needs to be raised. If we do not try to analyse and tackle problems early enough, we will be faced with increasing dilemmas.

NOTES

1. See the London Report of the Ten on improvements in Political Cooperation, 13 October 1981, part 1, German version in Europa-Archiv, 2/1982, p.D45-D50.

2. See especially the Final Declaration of the Paris summit 1972 and the Copenhagen Document on the European Identity from 1973, in Press and Information Office of the Federal Government (ed.), The European Political Cooperation, Bonn, 1978, pp.21 ff. and pp.69 ff.

3. Statement of the Conference of the Heads of State and Government of the Member States of the European Community, Paris, 21 October 1972, published in Press and Information Office of the Federal Government, op. cit., p. 44.

4. Commission of the European Communities, Commission Communication to the Council on a Mediterranean policy for the enlarged Community, Com (82) 353 final, Brussels, 24 June 1982, p.19. See also the memorandum on the development policy of the European Communities, Com (82) 640, Brussels, Oct. 1982, p.IV (also called Pisani memorandum).

5. Cf. the respective passages in the programs of the European parties in Programme fuer Europa. Die Programme der Europaeischen Parteibuende zur Europa-Wahl 1979, Bonn, pp.175, 308. The Economic and Social Committee of the European Community in its publications concentrates on the economic aspects of the 'global' approach and the repercussions of the EC enlargement; cf. Wirtschafts - und Sozialausschuss der Europaeischen Gemeinschaften, Die Aussenbeziehungen der EWG - Bilanz und Kohaerenz, Brussels, 1982, pp.59 ff.

6. European People's Party, European Digest 45."Study Meeting, 31 Aug. to 4 Sept. 1981, Naples", Luxembourg, 1981, p.228.

7. See Detlev Puhl, Die Mittelmeerpolitik der EG, PhD. Thesis, Tuebingen, 1981, to be published in 1983 in Kehl.

8. The treaties came into force on 1 Nov. 1962 (Greece) and on 1 Dec. 1964 (Turkey). That is why the years 1962 and 1964 are cited, too.

9. For more details see Heinz Andresen, "Ueber die Verwirklichung einer gemeinschaftlichen Mittelmeerpolitik", in Europa und die arabische Welt. Probleme und Perspektiven europaeischer Arabienpolitik, Bonn, 1975, pp.293-326.

10. Libya has been unwilling to negotiate with the EC so far, but there are signs of a changing attitude.

11. Press statement of the Heads of State and Government in Paris 1972, Press and Information Office of the Federal Government, op. cit., pp.5 ff.

12. See for an explanation of this notion the article by François Duchene, "The European Community and the uncertainties of interdependence", in Max Kohnstamm and Wolfgang Hager (eds.). A Nation Writ Large? Foreign Policy Problems before the European Community, London, 1973, pp.1-21. Some elements are reiterated in the Pisani memorandum, Com (82) 640, op. cit., p.31.

13. For details about its structure and developments see David Allen, Reinhard Rummel, Wolfgang Wessels (eds.), European Political Cooperation: Towards a foreign policy for Western Europe, London, 1982.

14. First Report on Political Cooperation, 27 October 1970 (Luxembourg Report), cited from Press and Information Office of the Federal Government, op. cit., p.29; Second Report on Political Cooperation, 23 July 1973 (Copenhagen Report), Ibid., pp.51-66; Third Report on Political Cooperation, 13 October 1981 (London Report), Europa-Archiv, 2/1982.

15. Udo Steinbach, "Die Europaeische Gemeinschaft und die arabischen Staaten", p.203, in Karl Kaiser and Udo Steinbach (eds.), Deutsch-arabische Beziehungen. Bestimmungs-faktoren und Probleme einer Neuorientierung, Munich, Vienna, 1981, pp.185-204. Wolfgang Wessels, "Die Europaeische Politische Zusammenarbeit", p.119, in Werner Weidenfeld and Wolfgang Wessels (eds.), Jahrbuch der Europaeischen Integration 1980, Bonn, 1981, pp.115-127.

16. Ibid., p.122. W. Wessels, "Die Europaeische Politische Zusammenarbeit", to appear in Werner Weidenfeld and Wolfgang Wessels (eds.), Jahrbuch der Europaeischen Integration 1981, Bonn, 1982, pp.293-308.

17. For more information, see Udo Steinbach, "The European Community and the United States in the Arab World: Political competition or partnership?", in Haim Shaked and Stamar Robinorid (eds.), The Middle East and the United States, New Brunswick, London, 1980, pp.125-135.

18. W. Wessels, "Die Europaeische Politische Zusammenarbeit 1981" op. cit.

19. U. Steinbach; "The European Community and the United States...", op. cit., pp.132 ff.

20. See for example Hélène Delorme and Marie-Claude l'Hyver, "La place de l'Europe des Neuf dans l'approvisionne-

ment agro-alimentaire du bassin méditerranéen: à la re-cherche d'une politique", in Colloque Cedece, La Communauté économique européenne élargie et la Méditerranée: quelle coopération, Presses Universitaires de France, Paris, 1982, p.205.

21. Deutsche Press-Agentur Die deutsch-arabischen Beziehungen, dpa-Dokumentation/HG 2988, Hamburg, 1981, p.20. Chancellor Helmut Schmidt, address to the German Bundestag, 8th session, meeting 203, 28 February 1980. "Preliminary formal contacts" between the EC and the Cooperation Council of the Gulf states (uniting Saudi Arabia, Oman, Qatar, Kuwait, United Arab Emirates) took place in June 1982. See Agence Europe, No. 3386, 9 June 1982.

22. Com (78) 200 final, Brussels, 27 April 1978.

23. See Commission of the European Communities, op. cit., p.1; Agence Europe, no. 1215, 23 July 1982.

24. Com (82) 640, op. cit.

25. Thus it was long discussed whether to include security questions on the agenda of the EPC; finally, in the London Report (op. cit., p.D46 of German version) it was agreed to include "political aspects of security" questions in the discussions among the Ten. The debate over the Genscher/Colombo initiative for a European Union has also revealed that there are strong divergencies over this matter.

26. See, for example, Wolfgang Hager, "Das Mittelmeer - 'Mare Nostrum' Europas?", in Max Kohnstamm and Wolfgang Hager, op. cit. (here the German version published in Frankfurt a.M. 1973), pp.233–264.

27. See Gunther van Well, "Mittelmeerpolitik und Suederweiterung", in Integration, 1/79, Bonn, 1979, pp.3–9; Lothar Ruehl, "Die Atlantische Allianz und die politische Stabilitaet in Suedeuropa", in Karl Kaiser and Karl Markuskeis (eds.), Sicherheitspolitik vor neuen Aufgaben, Bonn, 1979, pp.3–49.

28. Information on Europe and the Greek-Turkish dis-putes drawn from Roswitha Bourguignon, "Turkey and the EPC: balance and options", January 1983 (working paper for a conference on Turkey and the Community, organized by the Orient-Institut in Hamburg, Jan. 13–15, 1983).

29. Report of the Ten on improvements in Political Cooperation, Europa-Archiv, 2/1982.

30. See the concern of the Commission of European Communities, Com (82) 353 final, op. cit., p.9.

31. See the respective chapters in this book.

32. See the contributions of Gerald Blake and Geoffrey Marston in this volume; Vatroslav Vekaric, "Beispiel einer erfolgreichen Zusammenarbeit der Mittelmeerlander", in Internationale Politik, Vol. 33, 20 September 1982, pp.18 f.

33. See Patrick Daillier, "La Coopération dans la gestion de la Méditerranée", in Colloque CEDECE, op. cit., pp.354-392; EP Dok 1-949/82 Bericht Gautier ueber die gemeinsame Fischereipolitik fuer das Mittelmeer.

34. See Stefan A. Musto, "Die Zukunft der Mittelmeerpolitik der Europaeischen Gemeinschaft, Umlenkung oder Schaffung von Handlungsspielraumen", copied manuscript.

35. We are referring to Kissinger's Chicago speech of 1973 in which he said that Europe has only "regional" responsibilities whereas the U.S. has "worldwide" interests; see Department of State Bulletin, Vol. LXVIII, No. 1768, from 14 May 1973, pp.593-598.

36. See the Poettering report of the European Parliament, European Parliament Doc. 1-736/81. VWD Europa, 16 February 1982. Besides the definition of the new Mediterranean policy the Commission recently has outlined "Integrated Programmes" in favour of the Mediterranean regions of the Community; details in Agence Europe, No. 3388, 11 June 1982.

37. This proposal has been launched by Malta.

38. Helmut Hubel, Die sowjetische Nah- und Mittelostpolitik. Bestimmungsfaktoren und Ziele sowie Ansatzpunkte fuer Konfliktregelungen zwischen West und Ost, Bonn, 1982.

39. Compare Stefan A. Musto, "The EEC in search of a new Mediterranean policy", in C. Pinkele (ed.), Stability and Change in the Mediterranean, Praeger, New York, 1982.

40. See, for example, Ahmed Mahiou, "Note sur les rapports Algérie-CEE, ambiguités et paradoxes", in Colloque CEDECE, op. cit., pp.220-227.

41. See Sergio Minerbi, "The Accession of Spain to the EEC and its repercussions on the Israeli economy", Brussels, 1982 (copied manuscript), p.24, according to him Commissioner Pisani, responsible for the EC relations with the Mediterranean, thinks it could be a good idea.

42. See e.g. Stefan A. Musto, "The EEC in search of a new Mediterranean policy", op. cit.

43. Com (82) 353 final, op. cit., p.17.

44. See the notes above (especially note 36).

45. See especially Albert Bressand, "Monologues, dia-

logues, trilogues", in Colloque CEDECE, op. cit., pp.45-53; in 1979, the former French President Giscard d'Estaing launched the proposal of a summit conference of the members of the EC, the Arab League and the Organization of African Unity in order to establish closer links among these groups around the Mediterranean; see Frankfurter Allgemeine Zeitung, 26 May 1979.

46. As a first step in this direction, one could mention the cooperation between the Commission and the Arab Funds (fund of Saudi Arabia, Kuwait, Abu Dhabi, Qatar, OPEC, and others) concentrating on co-financing projects in the framework of Lomé I and II; see Agence Europe No.3383, 4 June 1982; No. 3387, 10 June 1982.

47. Alfred Tovias, "La politique méditerranéenne de la CEE face au dialogue euro-arabe et les plans de coopération avec les pays du Golfe: projets complémentaires ou rivaux?" in Colloque CEDECE, op. cit., pp.238-248, discusses a functionalist approach according to the different interests of the Arab Mediterranean countries compared with those of Cyprus, Israel or Yugoslavia. In his proposal, the field of financial and technological cooperation between the Ten and the Maghreb/Mashreq countries could be discussed in the framework of the EAd, whereas trade should be tackled with due respect to ACP rules (p.246).

48. See for this in general Stefan A. Musto, "The EEC in search of a new Mediterranean policy", op. cit., pp.3 ff.

49. According to François de la Serre, "The Community's Mediterranean Policy after the second enlargement", in Journal of Common Market Studies, No. 4, June 1981, pp.377-387, the present situation "does however compel Greece's partners within the Community to calm Turkish fears (perhaps by associating Ankara with the process of political cooperation?)" (p.384).

Chapter Nine

SOVIET STRATEGY AND THE OBJECTIVES OF THEIR NAVAL PRESENCE
IN THE MEDITERRANEAN

Robert G. Weinland

INTRODUCTION

This discussion has three objectives. Providing a
definitive description of the growth and current configuration of Soviet naval forces in the Mediterranean is not one
of them. The same applies to attempting to identify the
direct antecedents of specific actions those forces have
undertaken. Neither is feasible in a discussion of this
nature.

Neither is, however, essential for identifying the
general structure and content of the policies that have
guided the Soviets in establishing and exploiting politically their Mediterranean naval presence. And in view of
developments suggesting that these policies are in flux, it
seems unwise at this point to attempt more than that.
Sketching the broad outlines of those policies thus constitutes the first objective of this discussion.

Identifying the indications that those policies may be
changing (or may already have changed) constitutes its
second objective. Forecasting what these changes might bring
- in particular, what they portend for efforts to negotiate
restrictions on the presence and activities of superpower
forces in the Mediterranean - constitutes the third.

The discussion begins with a brief attempt to locate
naval strategy in the Soviet scheme of things. Next, it
addresses the evolving expectations of future war and prescriptions for its conduct that have structured Soviet military thinking over the last two decades, and consequently
seem likely to be reflected in the Mediterranean Squadron's

war plans. These expectations and prescriptions also shape the peacetime operations of the Squadron, the conceptual background of which is then discussed in some detail.

In dealing with both their planning for wartime and their policy in peacetime, the discussion presents a number of descriptions of Soviet strategy - i.e., the objectives they seek and the course(s) of action they would follow to achieve them. These descriptions are, of course, nothing more than inferences. In the case of Soviet planning for war, the inferences drawn have two origins: Soviet military doctrine (as reflected in their military literature), and the logic of the situation. They cannot be validated. In the case of Soviet policy in peacetime, the inferences drawn are amenable to validation. The actions they take can be examined, evidence can be marshalled, hypotheses can be tested.

The discussion consequently proceeds from postulations about Soviet strategy to observations about Soviet actions. And it is those observations and what they suggest about the strategies postulated that inform the attempt at the end to outline future Soviet policy and practice.

As should be apparent by now, this discussion lays no claim to certainty. Certainty cannot in any event be achieved. Soviet strategy per se remains hidden from view, and subject to change. Fragmentary evidence from statements and actions provides glimpses of it, but no more. Marshalled appropriately, those glimpses outline its general thrust, but no more. All that can be achieved is some reduction in our uncertainty about Soviet intent. That, one hopes, has been achieved.

NAVAL STRATEGY

Despite the protestations of some Soviet naval enthusiasts to the contrary, there is no such thing as Soviet "Naval Strategy". What in the West would be considered and treated as such is in the Soviet Union subsumed under the general rubric of "Military Strategy".

This is no idle distinction. It reflects the fact that the Soviet military establishment has been, is now, and in all probability will continue to be both highly integrated and dominated by the ground forces. As a result, in military affairs, the Soviet Navy is anything but an independent entity. It is one component of a larger whole, and the role envisaged for it in the direct defence of the Soviet Union and its Warsaw Pact allies is closely coordinated with (and indeed cannot be meaningfully considered in isolation from) the roles envisaged for the other branches of the Soviet armed forces (or for that matter, the forces of the other members of the Warsaw Pact).

In political-military affairs, on the other hand, where not the direct but the indirect defence of the Soviet Union and its Warsaw Pact allies and the protection and promotion of Soviet overseas interests are concerned, the situation is quite different. Here, the Soviet Navy appears to have achieved the status of a "senior service". Its importance as an active instrument of Soviet foreign policy and its capability to operate in that capacity as an independent entity are now clearly beyond question. This does not mean independent of Soviet political control, but independent of most of the remainder of the Soviet military establishment - the exception being the Strategic Rocket Forces, which, with a major assist from the Navy's strategic missile submarine component, provide the deterrent umbrella under which Soviet foreign affairs are conducted.

A unique system of views has been developed to structure the actions of the Soviet Navy in this latter, political-military, capacity. In this sense, there is a "Soviet Naval Strategy", but it is a political strategy, focussing on the peacetime rather than the wartime utilization of the fleet. In implementing this strategy, the main thing the Soviets are attempting to do is modify the behaviour of other actors in the international arena, and they are relying mainly on exploitation of the political influence potential of the forces they deploy outside their home waters in peacetime, rather than combat actions per se, to achieve those modifications.

PLANNING FOR WARTIME

What would the Soviet Mediterranean Squadron do in wartime? What would its combat objectives be; how would it set about their accomplishment?

In part, what the Squadron did would reflect the specific circumstances of the war: its antecedents and geographic focus, the strength and disposition of the forces available to each side, etc. In part, what it did would also reflect what the opposition did, especially if the opposition was able to seize the initiative. In part, therefore, what the Squadron would do cannot be predicted - or, more accurately, can only be forecast in terms of whatever specific conflict scenario is posited.

For the most part, however, at least in the opening phases of a conflict, Mediterranean Squadron operations would be structured by the Soviets' integrated combat plan. That plan in turn would reflect - be, in fact, a concrete expression of - Soviet military doctrine. This doctrine provides Soviet planners with a uniform system of expectations concerning the character of future war and dictates with regard both to how it should be fought and to what should be done to prepare for it. Those expectations and dictates are stated in the contract. Their impact is nonetheless pervasive. Insight into these prescriptions is consequently useful in forecasting aspects of the Squadron's operations that would be present in all scenarios.

Its "official" status and abstract character notwithstanding, Soviet military doctrine is not immutable. The expectations and prescriptions embodied in it change as the Soviets' definition of the situation and evaluation of their ability to cope with it change. Reviewing its evolution over the last 20 years outlines their current stance with reasonable clarity, and identifies those developmental trends and patterns in their perceptions and policies that seem most likely to persist. (1)

Two such trends are discernible in Soviet expectations of the character of a future war. One involves the degree of restraint expected to be exercised by the belligerents, which is seen to be increasing. The second - a concomitant of the first - involves the anticipated length of the conflict, which is also seen to be increasing. Both appear to have been incorporated into Soviet planning.

In the early 1960s, the Soviets held that conflict between the Superpowers would automatically escalate to all-out, world-wide, nuclear war. In the mid-1960s, they modified that forecast, dropping their contention that escalation necessarily would occur. In the early 1970s, they changed it again, concluding that, should war between the coalitions develop, although inevitably nuclear and world-wide, it need not necessarily be all-out. Intra-war deterrence was feasible. In the mid-1970s, they made a further modification. They concluded that, although inevitably nuclear, a war between the coalitions need not necessarily be world-wide. Expansion of the scope of conflict could be deterred. Now, in the early 1980s, they appear to have concluded that even a coalition war can remain conventional. Intensification of the level of conflict can be deterred as well.

In the early 1960s, they held that the inescapable escalation to all-out, world-wide, nuclear war would occur immediately. In the mid-1960s, as they began to recognize limitations on the intensity of conflict, they also began to recognize limitations on the dynamics of escalation. They began to plan for a "war by stages". Not only was such a war likely to begin at the conventional level, it could remain at that level for some time before escalating. Since then, the prospective length of that opening, conventional phase of the war has grown significantly (from, say, a week or two in the late 1960s-early 1970s to as much as, say, a month in the late 1970s. By the early 1980s, the war between the coalitions that they had begun to feel could be held at the conventional level was seen as likely to last for as long as three to six months.

Many of these changes in Soviet expectations were reflected in changes in their prescriptions for the employment of their naval forces. The two changes with greatest impact on Soviet naval presence and activities in the Mediterranean occurred in the mid-1960s, when they dropped their contention that conflict between the Superpowers would necessarily escalate, and in the mid-1970s, when they began to forsee conventional conflict between NATO and the Warsaw Pact lasting for an extended period.

The first of those modifications probably reflected their perception that, after a number of false starts and setbacks, they were finally on the way to acquiring a viable strategic deterrent, and as a consequence were acquiring increased freedom of action in the international arena. That made it possible for the Soviets to contemplate the exploitation of one of their principal assets (military power) in situations and for purposes (in peacetime, as an instrument of political influence) previously denied them. (2)

Confrontations between the Superpowers' military forces obviously could produce conflict. As long as the Soviets perceived such conflict - no matter what its scope or level of intensity – as no less than a preliminary to all-out war between the coalitions, the risks of such confrontation in situations where central values were not threatened were unacceptably high. As soon as confrontation-produced conflict between the Superpowers was perceived as no more than a preliminary to all-out war – which consequently could be avoided – then the range of situations in which such confrontations could be staged with acceptable risk expanded to encompass the protection of less-than-central values.

As will be outlined further below, this doctrinal change not only "permitted" the establishment of a militarily significant Soviet presence in the Mediterranean in the mid-1960s, (3) and the subsequent employment of those forces for politically significant purposes, but accorded a degree of priority to both. The second doctrinal modification referred to above, which came a decade later, appears to have shifted Soviet priorities elsewhere, and may have had the effect of establishing limits on their Mediterranean presence and its exploitation.

As long as it was felt that any conflict that erupted between the superpowers could not be constrained, it was imperative that no action be taken that could lead to its initiation. Once it was recognized that escalation to all-out war was not inevitable, that conflict could be constrained, it became imperative that action be taken to impose such constraints. Posing a deterrent to conflict-engendering undertakings by the other side represented one such action. Preemptive strikes to eliminate the other side's ability to expand the scope or increase the intensity of a conflict represented another. From the beginning, the mission of the Soviet Mediterranean Squadron appears to have encompassed both. It apparently still does.

It also appears, however, that neither mission is now accorded the relative priority it once had. After it was recognized that conflict between the superpowers not only could be constrained, but probably would be – inherent in the recognition that conflict was likely to begin at the conventional level, and could remain there – the Soviet's action imperatives were modified. Once again, this shift in prescription probably reflected a shift in their perception of the strategic balance: by the mid-1970s, the Soviets felt that they had acquired an assured destruction capability

272

like that they knew the United States possessed. As long as that capability for assured destruction remained mutual, neither side had an incentive to escalate a conflict to the point where those capabilities would be employed.

The Soviets' action imperatives under these conditions were substantially different. The scope and impact - and, most likely also pace - of a conflict fought with strategic weapons could, and hence probably would, be great. That was unlikely to be the case in a conflict fought <u>without</u> strategic weapons. The range and destructiveness of tactical weapons were, by definition, far less. The pace of a conflict fought with such weapons would be, of necessity, much slower.

In a short war to end in strategic interchange, priority had to be accorded to mounting preemptive strikes against the strategic offensive capabilities of the other side that could be attacked in time. The Mediterranean Squadron's well-documented anti-carrier mission, and the steps the Soviets took to maintain the ability to mount such strikes, both reflect just such an imperative.

In a longer war, particularly in a war in which there is reason to believe that strategic weapons may not be used at all - or, if so, then only long after the initiation of conflict - other threats and opportunities emerge, and a different constellation of actions must be accorded priority. (4) For example, as long as the Soviets expect to be able to bring a conflict in Europe to a successful conclusion in days or at most weeks, they do not need to allocate forces to stopping the movement of men and material from North America to Europe. Although such movements could determine the outcome of the war, they are unlikely to occur. It would take at least weeks, perhaps months, to execute them. However, if European NATO can contain a Soviet advance, creating an opportunity for the movement of reinforcements and resupply from America to take place, that advance can be repulsed. As a result, the longer the Soviets consider a conflict on the Central Front to be likely to last, the more importance they must accord the interdiction of NATO's trans-Atlantic lines of communication. If they expect the war to go on for months, and wish to win, they must allocate forces to the task of interdicting those lines of communication.

In similar fashion, as long as the Soviets can count

on a potential war's being "nasty, brutish, and short", the question of protecting their sea-based strategic offensive forces - a principal component of their assured destruction capability - is not likely to arise. Strategic offensive forces were taken to sea so that they would be invulnerable to preemptive strike by opposing strategic offensive forces. They remain so. NATO, however, has impressive tactical sea-control capabilities. In a short war, those tactical capabilities pose little threat to the Soviets' SLBM force. A long war, on the other hand, would provide an opportunity for those tactical capabilities to be employed strategically. As a result, the longer the Soviets consider a conflict to be likely to last the more importance they must accord to providing direct protection to their SLBM force.

Both of these missions - interdicting NATO's lines of communication across the Atlantic and providing for the tactical defense of their sea-based strategic offensive forces - require roughly the same type of submarines and surface combatants the Soviets have been deploying to the Mediterranean since the mid-1960s. Neither mission, however, can be carried out effectively by forces located there.

Competition for employable resources is a predictable consequence of such a shift in Soviet priorities. As will be indicated below, the functions the Soviets have long felt it necessary to have performed in the Mediterranean seem not to have won out in that competition.

This should not be taken as an indication that those functions have lost their importance in any absolute sense (although recent Soviet behaviour suggests this could be the case). It only indicates that other functions and other regions have acquired increased importance in Soviet planning.

POLICY IN PEACETIME (5)

Navies are, first and foremost, warfighting instruments. But they are also useful, and used, in peacetime.

There are two fundamental categories of reasons why a state would use its warfighting instruments in peacetime. One is to implement its war-related policies (for example, attempting to deter war, or insuring readiness to fight should war occur). The other is to implement those of its policies that are not war-related (for example, supporting foreign policy or protecting overseas interests).

Even these very broad distinctions (not to mention the possible sub-divisions of each category) can be rendered academic in practice, since a single action can serve more than one end. Moreover, they are not of equal importance. They do, however, provide a simple standard, based on familiar concepts and logic, against which to compare Soviet statements and actions, some of which reflect modes of thinking that are, to say the least, unfamiliar.

How does the Soviet employment of its naval forces in peacetime compare with this basic breakdown of functions? Several observations can be made straightaway.

First, as far as the Soviets are concerned, the two war-related functions of deterrence and preparation for combat are closely linked. They consider the achievement of the capability to fight a war successfully (as opposed to being prepared only to punish a potential attacker) as being a most important factor - some would say the most important factor - in the deterrence of war. Thus, despite the importance they attach to avoiding armed conflict, much of their peacetime naval activity is devoted to direct preparation to fight. Second, they consider deterrence of war, and preparation to fight, to be important ends not only of their military activity but also their foreign policy. Thus, given the active role assigned the Navy in direct support of Soviet foreign policy, and the active role played by Soviet foreign policy in advancing Soviet security, much of their peacetime naval activity is devoted to indirect preparations to fight as well. Consequently, the first three of the peacetime functions noted above - deterrence, preparation to fight, and direct support of foreign policy - are to some degree indistinguishable in Soviet eyes, and hard for Western eyes to differentiate in Soviet practice.

This is all somewhat abstract. The urge to move from the abstract to the concrete in search of clarity should be resisted, however. It won't necessarily improve our understanding of Soviet behaviour, since some of its policy antecedents appear to be abstract in the extreme.

Two political functions carried out by the Soviet Navy in peacetime will be discussed in depth here. The first of these two functions, "active defence of peace and progress", represents an attempt to achieve two no less abstract ends: support of "progressive change" and prevention of war. The objective of the second function, "preparation of maritime theatres of military operations", is more practical: improving the likelihood of success should combat take place. Both are carried out by the same means: the manipulation of naval forces to influence the behaviour of other actors in the international arena. They differ only in the ends they serve. Both characterize Soviet Naval activity in the Mediterranean.

Active Defence of Peace and Progress

If the Soviets actually view the international system in the terms in which they describe it - and this is admittedly a big "if" - then a substantial portion of their naval activity in the Mediterranean (and elsewhere in the Third World) may be devoted to an attempt to intervene in, and alter, what they consider to be the "normal" progression of events in international conflict. The objectives of such intervention are not only to create and maintain a favourable political-military environment for what they refer to as "progressive change", but also to reduce the likelihood of their becoming involved in a major war triggered by an attempt to effect such change. Their descriptions of the way the international system works, and in particular the processes of international conflict, are tortuous and reflect distortions in perception and reasoning that can be traced directly to Marxist-Leninist ideology. However, these descriptions are not completely divorced from reality; and following the Soviets' basic argument from premises to conclusions provides potentially useful insight into some of the considerations that may be motivating them.

Defence of Peace

There are two ways in which the Soviet Armed Forces are considered to serve, albeit indirectly, in the defence of the homeland in peacetime. The first is, of course, through strategic deterrence. The second, and for this discussion more interesting, way is through what might be termed "local deterrence": deterrence of the reactionaries' and imperialists' use of their military forces, first to start local conflicts in the Third World, and then to attempt to influence the course and determine the outcomes of those conflicts. The Soviet Navy is held to play a leading role in both types of deterrence.

The Soviets' perceived requirement for local deterrence is a direct outgrowth of the way in which they view the international situation - in particular, the situation in the Third World. As they see it, the most important characteristic of contemporary developments there is the continuing, historically-determined process of what they refer to as "progressive" change. At the domestic level they see this process producing radical political reorganization and socio-economic transformation (following the "socialist" example). On the international level, they see it as leading to the establishment of what they consider to be national political and economic independence (by which they mean independence from the capitalist-imperialist West). The principal protagonists of "progressive" change are what they term the "progressive" forces: national liberation movements, newly-independent states, and, of course, world socialism.

The principal antagonists to "progressive" change are the aggressive forces of local reaction (they would cite Israel as an example) and world-wide imperialism (led, of course, by the United States). Attempts by reactionaries and imperialists to stop progress, and eventually reverse it, both cause and exacerbate local conflicts.

Local conflicts occur frequently in the Third World. The Soviets consider them to have two causes. One is a local contradiction (economic, political, military, ideological, territorial, national, ethnic, etc.). The other and more prevalent cause of local conflict is held to be the aggressive actions of the forces of reaction and imperialism. The most frequent targets of these local aggressive actions are the "progressive" forces - national liberation movements in particular.

In the Soviet view, regardless of how local crises originate, the aggressive forces almost invariably intervene in them in order to advance their own interests. These interventions result in the escalation and expansion of these conflicts, producing threats not only to regional but to world peace as well.

Given that they see the world this way, the Soviets see an imperative to pursue, in concert with other forces of "peace and progress", two objectives. One is to protect and promote "progressive" change. The other is to prevent the exacerbation of local conflicts. Both objectives are served by "actively counteracting" the attempts of "aggressive" forces to start and exploit local conflicts.

"Active counteraction to imperialist aggression" is called for because of what the Soviets consider the pernicious effects of the involvement of imperialist powers in local conflicts. Their intervention not only delays future progress, but by threatening regional and world peace, threatens progress that has already been achieved – the "gains of socialism". If unchecked, the imperialists' proclivity for intervention in local conflicts could eventually create a situation placing not only progress per se but the security of the Soviet homeland (the bastion of the forces of "peace and progress") in jeopardy.

"Actively counteracting imperialist aggression" could, however, also prove dangerous for the Soviet Union. The involvement of the great powers of <u>both</u> the imperialist and socialist camps in a local conflict could transform such a conflict into what the Soviets refer to as an acute international political crisis. And that developments, because of the perceived propensity of the United States to threaten the use of all its forces (i.e., from local conventional to strategic nuclear) in carrying out its "from a position of strength" crisis management policy, could lead to global nuclear war.

Fortunately, from the Soviet point of view, the correlation of forces between the two camps, which the Soviets see as having over the last decade increasingly come to favour the forces of "peace and progress" over those of reaction and imperialism, has had an inhibiting effect on the aggressors. This shift in the correlation of forces, and the continuing implementation of the Peace Programme first promulgated by the XXIV and endorsed by the XXV and XXVI CPSU Congresses, are bringing about a radical restructuring of international relations.

The Peace Programme calls for the Soviets to undertake three closely related action programmes. The first is a fundamental reorganization of the international political-military environment (through negotiations, agreements, etc.). The second is the modification of critical interstate relations (the consolidation of détente with the West and enhanced collaboration with the other members of the Socialist Community). The third is "active counteraction to imperialist aggression".

The last of these, which provides the content of the

"international mission" of the Soviet Armed Forces, foresees
the performance of two separate tasks: "stopping aggression", and "supporting victims of aggression". "Stopping
aggression" involves preventing acute international political crises from occurring (by deterring both the aggressive actions of reactionary forces that cause local conflicts and the imperialist interventions that exacerbate
them). It also involves regulating those crises that can't
be prevented (by deterring both the imperialists' threats to
use their nuclear forces and their demonstrative movement,
concentration and actual use of their conventional forces).
It is these particular functions - intended to control both
the initiation and the continuation of local conflicts, and
prevent their evolution into major war - that forward-
deployed Soviet naval forces are performing when engaged in
the "active defence of peace".

"Supporting victims of aggression", the second of the
two tasks carried out in actively counteracting imperialist
aggression, can involve the provision of military assistance
(including even direct support and combat forces) to national liberation movements and newly-independent states. Since
this activity is oriented not toward Soviet self-defence but
toward the protection and promotion of Soviet overseas interests, and since there are conclusions to be drawn regarding
the defensive purposes of Soviet forward deployments, discussion of the use of the Navy in support of "progressive
change" will be delayed for a moment.

Assuming the depiction of the Soviet perspective on
international conflict outlined above to be more accurate
than not, and assuming that perspective to be more influential than not in Soviet decision-making, the deployment of
Soviet naval forces first into the Mediterranean and subsequently to other areas of the Third World could have had an
important purpose, perhaps too readily discounted in the
West as reflecting only empty rhetoric. That purpose need
not have been to oppose the aggressive actions of local
reactionary forces, or for that matter support the actions
of the progressive forces, although elements of both have
unquestionably been present. It might have been to at least
constrain if not actually prevent intervention in local
conflicts by the great powers of the imperialist camp. Most
importantly, it might have been to deter the United States
from threatening, if not actually employing, its conven-

tional area control and projection forces to determine the outcomes of such conflicts (in particular, to deter the United States from further such use of its Sixth Fleet). One reason for the Soviets adopting such a dangerous course might have been to avoid something they considered even more dangerous: to preclude the "necessity" for Soviet counter-intervention, with its unpredictable but potentially even more explosive consequences, and thus keep general war at as great a distance from the Soviet Union as possible.

That perspective on conflict has several potentially important implications for the Soviets in their approach to involvement in peripheral conflicts. First, local conflicts (at least those growing out of local contradictions) are not considered to represent a threat to regional or world peace. Second, the involvement of "peace-loving" forces in local conflicts, including the involvement of the Soviet Union or the other major powers of the Socialist Community, is not considered to represent a threat to regional or world peace either. It is only when the major powers of the imperialist camp intervene that such a threat emerges. Third, it is only when the great powers of both systems become involved that the threat of world nuclear war arises.

In other words, using the October 1973 Arab-Israeli War as an example, the Arabs could attack the Israelis, and the Soviets could assist the Arabs in preparing for the attack and sustaining the fight without the emergence of a threat to other than local peace. But as soon as the United States began to assist Israeli activity, the situation threatened to get out of hand - hence the Soviet emphasis on deterring direct U.S. intervention, and their reluctance to go very far with their own movement toward direct intervention.

Defence of Progress

If accurate, and influential in Soviet policy-formulation, this definition of the international situation and the policies and practices required to cope with it may explain some of the things the Soviets have (and have not) done with their Navy, but not all of them. Both common sense and careful analysis of Soviet behaviour suggest that the Soviet naval presence in the forward area in peacetime is intended to do more than just defend peace. That's a negative objective. They also have some positive objectives: things they would like to see happen, particularly in the Third World.

What is the Soviet Union attempting to accomplish in the Third World, and how does it employ its military forces in the effort? Neither answer should be considered a mystery. They have told us at length what they intend, and we have seen in detail what they do.

As indicated further above, they are attempting to advance what they view as "progressive" change in the Third World. It was noted that they view local reactionaries and the forces of worldwide imperialism as the principal opponents of "progressive" change. And while such change can be delayed by reactionary/imperialist opposition, it is in the "progressive" direction that the Soviets see history moving.

They define the situation and structure their behaviour in the Third World in terms of that movement. For the Soviets, "progressive" change represents the status quo in contemporary Third World affairs. Their preferred role in those affairs is the establishment of "favourable conditions" for such change, and, in addition to the protection and promotion of their more prosaic interests (like insuring the safety of Soviet citizens and protecting trade flows) it is for that express purpose - establishing favourable political-military conditions for "progressive" change - that Soviet military forces, almost exclusively their naval forces, are employed for positive ends in the Third World. The mission of those forces is to defend the status quo. They perform that mission by deterring the initiations and compelling the cessation of what they see as attempts to alter that status quo: reactionary and imperialist efforts to stop or reverse "progress".

Soviet military forces consequently are not intended to be, and do not act as, the engines of "progressive" change in the Third World. They are not employed to overthrow established regimes; they do not participate in consolidating the gains of revolutions. Those functions are performed by other elements of the forces of "progress": national liberation movements, newly-independent states, and other instruments of world socialism - like the Cubans and East Germans. The Soviet military is simply "riding shotgun" for them, their efforts, and their achievements.

As such, the immediate objective of stationing Soviet forces in the Third World in defense of "progress" is once again deterrence, not war-fighting. The ability to fight effectively is, of course, a prerequisite for deterring effectively; but significant combat is not what they have in mind, or prepare for. Local reactionary forces are unlikely themselves to possess great military strength and therefore can be intimidated by the presence the Soviets maintain deployed forward. And when the far more numerous and capable forces of worldwide imperialism become involved - in particular, when the United States moves forces to the scene - the Soviets can deploy additional forces of their own to reestablish the deterrent counterweight of their presence.

The Soviet Union doesn't appear to possess either the combat forces or the support infrastructure that would be required to carry out such a mission if it entailed taking vigorous or forceful action in the Third World - especially if the ability to sustain high-intensity combat operations in distant areas were one of the requisites. In most circumstances, however, establishing "favourable political-military conditions" for "progressive" change is not that demanding. And should circumstances prove otherwise, the Soviets have demonstrated a remarkable ability to distance themselves from such situations. They are, after all, not themselves responsible for actually bringing about "progressive" change. Theirs is "a more lofty task..."

Preparation of Maritime Theatres of Military Operations

There are a variety of very practical reasons to move forces into potential combat zones in peacetime. Some, like being in the optimum position to fight now if necessary and improving one's capability to fight at some unspecified point in the future, need no elucidation. Both of these probably explain a significant portion of Soviet naval activity outside their home waters today. This arguably has been the case with their presence in the Mediterranean.

The Soviets have an additional reason to deploy their forces in potential combat zones in peacetime. It is oriented toward the same ultimate military end: structuring the situation to improve the likelihood that Soviet forces will prevail if war should come. But it employs a different means of achieving that end. Where the first two conflict-oriented rationales (establishing optimum position and improving readiness) involve taking military actions intended to enhance their own combat potential, this third rationale involves taking military-political actions intended to detract from their likely opponents' combat potential.

The political process involved is relatively straight-forward. It is commonly referred to as intimidation. It involves manipulating the peacetime presence and activity of Soviet forces in potential combat theatres (such as the Eastern Mediterranean) to affect to Soviet advantage the definitions of the situation and consequent policies of those they perceive likely to oppose them there. Specifically, this means taking actions that increase the likelihood that potential opponents will perceive the balance of military power in the region as lying so far in favour of the Soviet Union that it would not be cost-effective to attempt to challenge them there.

Along with other, more concrete measures, such as making preparations to control strategic locations, establishing land-based support facilities, providing for surveillance and setting up appropriate command-control and communications relationships, the Soviets obviously consider this political campaign an effective peacetime contribution to the establishment of favourable military conditions for winning a dominant position in a theatre in wartime. This approach to the problem of preparing to fight probably goes a long way toward explaining the frequency, location, magnitude, and openly demonstrative character of some of their major fleet exercises, and in particular the large-scale manoeuvres such as Okean and Vesna that they held in the 1970s. It also may help to explain their exaggerated efforts to defend the legitimacy of expanding their naval operations beyond Soviet home waters, and the vigour with which they tend to react to the presence of the forces of potential opponents in close proximity to the Soviet Union.

In essence, they devote a significant amount of effort in peacetime to creating an image of themselves as possessing overwhelming warfighting strength, not only in their home waters, where no one would doubt that, but in potential combat theatres in the forward area as well, where such doubts might be legitimate. They view this as a means of reducing the level of effort that must be devoted in wartime to the establishment and maintenance of control of those theatres - a task they recognize as a sine qua non of the successful performance of other, critically important, wartime functions.

THE EVOLUTION OF THE SOVIET NAVAL PRESENCE IN THE
MEDITERRANEAN

Soviet deployments in the Mediterranean have undergone
significant change since their inception. Both the size and
the composition of the force they maintain there have been
altered. These changes reflect - and illustrate - the evolution of Soviet expectations of future war and its imperatives outlined above. They also have significant implications for the exploitation of their presence for political
purposes in peacetime, which will be discussed further below.
It is not necessary to examine Soviet naval operations
in the Mediterranean in detail to recognize the most important of these changes (which is fortunate, since a substantial portion of the information required for such an examination has yet to be placed in the public record). Figure 9.1
describes the evolution of the Soviet naval presence in the
Mediterranean. Figure 9.2, which describes the evolution of
the world-wide Soviet naval presence during the same period,
provides an appropriate context for evaluating those Mediterranean deployments. Both illustrations are drawn in terms
of annual ship days. This is an aggregate measure of the
time spent, in the one case in the Mediterranean and in the
other outside home waters, by Soviet naval and naval-
associated units of all types. It is a reasonable represen-
tation of level of effort.
As figure 9.1 demonstrates, the evolution of the Soviet Mediterranean Squadron has progressed through three more
or less distinct stages. From their inception in 1964
through 1971, Soviet naval deployments to the Mediterranean
increased in scale steadily (from 1,500 ship days in 1964 to
19,000 in 1971). From 1972 through 1976, they fluctuated
(averaging 19,400 for the period, and reaching an all-time
high of 20,600 in 1973). In 1977, they fell back to roughly
16,500, and have remained remarkably close to that figure
ever since. As noted above, it looks as though a limit has
been placed on the presence the Soviets are willing (or,
perhaps, able) to maintain in the Mediterranean. When that
decision was taken is not altogether clear, although the
1976-77 period is a logical candidate. That some such decision was taken is, however, obvious.
As figure 9.2 demonstrates, worldwide Soviet naval
operations have also evolved in identifiable stages. In the

284

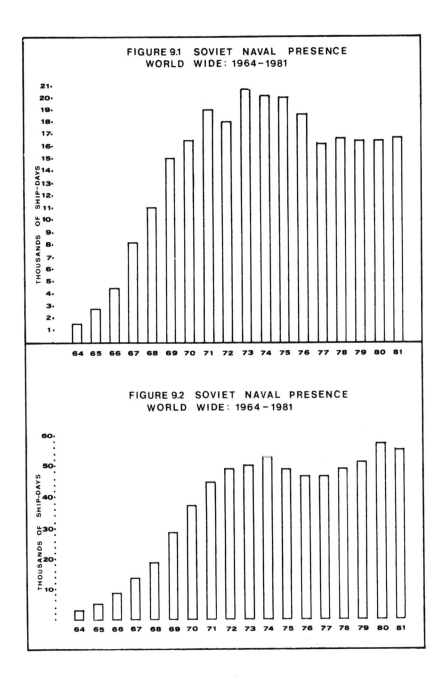

early years, the worldwide pattern (steady increase) parall-
eled that evident in the Mediterranean. This is not surprising. For the first five of those years, operations in the
Mediterranean accounted for more than half of the worldwide
total. In the most recent five-year period, the worldwide
pattern (renewed increase) has <u>not</u> been reflected in the
Mediterranean. This is surprising - but for other reasons,
since Mediterranean operations now account for less than one
third of the total.

Comparing figures 9.1 and 9.2 makes a second important
point about Soviet deployments in the Mediterranean: the
limit that has been placed on their presence there is
intentional. The cut-back in Mediterranean activity in the
second half of the 1970s could have been dictated by operational conditions. Throughout the first half of the 1970s,
the Soviets employed their naval forces not only extensively
but intensively. It is conceivable that this level of effort
was beyond what they could sustain, that increased deployments were undertaken at the expense of future availability
- for example, keeping ships in service by delaying over-
hauls. If that is what they did, and the cut-back in
Mediterranean activity occurred because the past caught up
with them, then the significance of that cut-back for explaining present and forecasting future Soviet actions in the
Mediterranean is diminished significantly.

However, the parallel cut-back in worldwide activity
that began in 1975, and could itself have been an artifact
of the situation just described, has now been reversed. The
total level of effort reflected in Soviet naval operations
undertaken in both 1980 and 1981 exceeds that achieved in
1974. Clearly, if in the second half of the 1970s the
Soviets were unable to achieve what they had in the first
half, they can surpass it now. And clearly, even if diminished operational availability were the explanation for the
reduction in Mediterranean operations in the second half of
the 1970s, (6) it does not explain their failure to increase
in parallel with increased operations elsewhere in the
1980s. They could have. They didn't. They clearly weren't
meant to.

Why this is so is not as clear. A rationale was
outlined further above, and the subject will be addressed
again further below. One observation should suffice for the
moment. The Soviets' decision to limit the size of the naval

presence they maintain in the Mediterranean obviously was accompanied (if not occasioned) by a shift in their priori- ties - possibly to the performance of other functions, but certainly to increased presence in other regions. (7)

This line of argument should not be taken very far. Soviet priorities may have shifted. Their resources may now be allocated differently. Even if they have, these changes do not alter the fundamental situation in the Mediterranean. The Soviet Mediterranean Squadron remains where and what it was. It has not been moved elsewhere. It has not been disbanded. It has not been disarmed. Its strength may be diminished somewhat, but it is still significant: the annual coverage for 1981 was some 45 units, both combatant and auxiliary. And it still possesses formidable combat capabi- lities. As before, it is what the Soviets want it to be: a force that cannot be ignored.

THE FUTURE

Any attempt to make a direct forecast of the future of the Mediterranean Squadron that does not acknowledge the evidence of its most recent employment is liable to be in error, particularly when that employment may represent a critical test of the utility the Soviets now attach to the maintenance of a naval presence in the region in peacetime and its exploitation for political purposes. This discussion attempts no such direct forecast.

It is possible, however, to approach the question of the Squadron's future indirectly: by addressing the factors that appear likely to shape its future. Two of these have been discussed at length above. One is the doctrinal expect- ations and prescriptions that appear to structure Soviet military planning for war. The other is Soviet perceptions of the dynamics of international conflict and the role in its "management" their forces should, and should not, play in peacetime. These two factors are not independent. The former drives the latter. What the Soviets do with their forces in peacetime is predicated on what they expect to have to do with them in wartime.

As outlined above, they appear to expect the United States and NATO - and themselves - to exercise increasing restraint in the use of force. And they appear to expect that restraint to translate into at least lengthier, if not less intense, conflict between the coalitions.

As their expectations of potential conflict change, their prescriptions for its conduct and the steps they must take to prepare for it change accordingly. Protracted conventional war appears more and more to be what they expect, and its requirements differ substantially from those that prevailed when the Mediterranean Squadron was being established. New requirements imply the reallocation of priorities among combat functions and theatres, and that implies the reallocation of resources.

It is possible that the apparent changes in the strength and activities of the squadron noted above reflect such reallocations. The importance of some of the functions previously performed by the Mediterranean Squadron may have been downgraded; resources previously deployed to the Mediterranean may have been assigned to other theatres to perform functions assuming increased importance.

If, in fact, this is what has occurred, it could be an indication that the role assigned to the Mediterranean Squadron in Soviet war plans has been downgraded. And that, in turn, could be an indication that attempts to negotiate limitations and eventual reductions in the presence and activities of Soviet forces in the Mediterranean might prove successful. But that isn't necessarily what has occurred, and even a significant downgrading of the combat role of the Mediterranean Squadron wouldn't guarantee the success of such negotiations.

First, the changes observed in the strength and activities of the Squadron could well be more apparent than real. Moreover, even if real, they do not <u>necessarily</u> imply a downgrading of combat functions previously assigned to the Squadron.

The unit capabilities of Soviet combatants have increased significantly over the years as new weapons and sensors have been developed and deployed to the fleet. The Soviets may have concluded that a lesser number of more capable units can still perform the Squadron's combat functions. In the same vein, some of the combat functions previously assigned to the Squadron could have been reassigned to other forces - the strike aircraft of the Black Sea Fleet, for example.

Second, even if the Squadron's potential wartime contribution were now so minimal as to permit its existence to be negotiated away, the continued performance of its peacetime functions could be considered of sufficient importance to justify its continued existence. Unlike many of its potential wartime functions, which could be performed – perhaps less efficiently – in other ways or by other means, the performance of the Squadron's peacetime functions requires its presence in the Mediterranean.

An assessment of the Soviets' current use of the Mediterranean Squadron as an instrument of international political influence is required to go beyond this point. If the Squadron's peacetime functions remain essentially what they were as recently as 1976, and if the Soviets still accord the performance of those functions the importance they accorded it then, it seems unlikely that they would agree to the Squadron's withdrawal from the Mediterranean – as long as local conflict remained endemic to the region, and the United States had not agreed to withdraw the Sixth Fleet. Even if those functions have been reduced in scope and importance, it seems unlikely they would withdraw unless those two additional conditions had been met.

NOTES

This paper is a personal, not an official interpretation. As such, it does not necessarily reflect the views of the Center for Naval Analyses, the U.S. Navy, or any other component of the U.S. Government.

1. Where this discussion treats questions of the evolution of Soviet military doctrine, it draws heavily on the analyses of the author's colleague, James M. McConnell - who, one prays, will be held blameless for any distortions it may contain.

2. Again, where this discussion treats questions of the conditions, objectives and limits of the political employment of Soviet naval forces, it is in many respects the product of sustained interchange with the author's colleagues, most notably Bradford Dismukes and James M. McConnell. Their interpretations and the evidence supporting them are detailed in: Dismukes and McConnell (eds.), Soviet Naval Diplomacy, Praeger, New York, 1975.

3. Actually, it would be more appropriate to refer to it as a "counter-presence", since the U.S. Sixth Fleet had already been in the Mediterranean for a decade and a half. And it might be more appropriate to refer to the "reestablishment" of a Soviet military presence there, since a contingent of Soviet submarines had been stationed in Albania from 1958 to 1961 (the basing arrangement became a casualty of the Sino-Soviet split). However, the military significance of this contingent was questionable, and it had little if any political impact. It is consequently ignored in this discussion.

4. For a discussion of this attention and resource allocation problem and some of its ramifications, see: Robert G. Weinland, Northern Waters: Their Strategic Significance, CNA Professional Paper No. 328, December 1980.

5. The argument advanced concerning the conceptual antecedents, structure and content of the Soviet Navy's peacetime mission was first presented, in a significantly different context, in: Quester (ed.), Navies and Arms Control, Praeger, New York, 1980.

6. An issue on which this discussion takes no stand. The author has argued elsewhere that loss of access to support facilities in Egypt contributed significantly to the

reduction. That is, however, a short-run phenomenon. Why in the long run (and surely, six years qualifies as the long run) the Soviets have not taken steps to rebuild their presence is a different issue. See: "Land Support for Naval Forces: Egypt and the Soviet Eskadra 1962-1976", <u>Survival</u> 20-2 (Mar/Apr 1978), pp. 73-79.

7. Primarily the Pacific and Indian Oceans (where, one should not forget, the United States has recently shifted some of the forces it previously kept in the Mediterranean).

Chapter Ten

AMERICAN FOREIGN POLICY, NATO IN THE MEDITERRANEAN, AND THE
DEFENCE OF THE GULF

Ciro Elliott Zoppo

EAST-WEST SECURITY AND NORTH-SOUTH POLITICS: A CRUCIBLE FOR
INTERDEPENDENCE

Three major factors have driven recent transformations
in the international system: the political use of oil
resources; new developments in military technology with
strategic significance; and the increasing interpenetration
of internal and external political systems.

The emergence of the geopolitics of oil has added an
economic dimension to national security, altering radically
the relations between the industrialized nations and
resource-rich countries of the Third World.

Developments in nuclear and conventional military
technology have created conditions that could effectively
destabilize the strategic relationship between the United
States and the Soviet Union, the Eurostrategic balance, and
the military gap between major powers and regional powers.
Thus, they threaten to erase the boundary between the superpower deterrent system and regional conflict systems. The
relative decline of American military capability vis-à-vis
the Soviet Union, and of the credibility of U.S. military
resolve everywhere, has intensified the effects of this
change.

Finally, the political turmoil in the Third World
resulting from the difficulties surrounding the transition
from colonial status to independent and viable nations has
become internationalized. This internationalization has
taken place not only in terms of superpower involvement but
also by the creation of regional conflicts that virtually

292

erase the distinction between internal and external politics
engendering an intensification of social and political instability to almost unmanageable levels of ferment.

The confluence of these factors into a dynamically
conflictual pattern has become most pronounced in the countries around the Gulf, a part of the Middle East and an
extension of the Mediterranean region. Because the Mediterranean's geography makes it the political, economic and military junction of Europe, Asia and Africa, the defence of the
Gulf is crucially linked with the military and political
assets of the Mediterranean.

The strategic equation between East and West in the
Mediterranean has all but disappeared. Inherently unstable
because of the perennial Arab-Israeli conflicts, the
East-West security dimension in the region has become further diffused by the additional political fragmentation
brought about by radical change in the political status of
Afghanistan, Iran and Ethiopia, the outbreak of the
Iraqi-Iranian war, and the ever-recurring cycle of
sub-national and state terrorism. Terrorism has by now
forged direct links between the Mediterranean and the adjacent Middle East and Gulf regions. (1)

In tandem with the political changes in the Gulf area
which have undercut the barriers to the spreading of local
conflict from the Gulf and the Middle East to the Mediterranean, developments in military technology have tended to
erase regional boundaries because of the changed operational
interactions between land-based air power and naval deployments. The increase in sophisticated air and naval capabilities of riparian, non-NATO, states have further undermined
the possibilities for a viable strategic equation and the
security of the Mediterranean.

At the level of regional conflict especially, increasing economic interest in Mediterranean resources combined
with disparate and sometimes conflicting political orientations have made the concepts and practice of balance of
power mechanisms basically inappropriate. There is little
consensus among the riparian countries, including members of
NATO, on the nature of the threats to their national security and the stability of the Mediterranean region.Greece,
for example, regards Turkey, its neighbouring member of
NATO, as the primary threat to its security. There is
similarly no consensus among them about the appropriate role

for the United States in the maintenance of security in the extended region from the Mediterranean to the Gulf.

This lack of consensus on the nature of the threats to the security of the region deprives the West of a coherent military strategy that could isolate, effectively, regional North-South conflicts from the potential for East-West confrontations. Without such a strategy, the possibilities for the development of politically viable economic agreements to exploit the resources of the region for mutual benefit are likely to be curtailed. Thus, in the extended Mediterranean region economic interdependence and security are intimately related for all nations of the area.

So long as the United States was the major and unchallenged military and economic power in the Mediterranean, there existed a strategic coherence and a stability in the security of the area. It was also possible for the United States to discharge the function of political mediator and local conflict and crisis manager. The United States, as the major Western military power in the region and the most influential politically, remains, in principle, the most capable Western power with the potential for integrating the defence of the area, and for arbitrating regional political disputes. However, trends seem to be favouring chaos over concurrence in the relations among Mediterranean countries and between the United States and its NATO allies. The nexus between economic interdependence and security may indeed be tested for Europe and the United States in the Mediterranean.

Even without a consensus among local states on the role of the United States in the Mediterranean and the adjacent Middle East and Gulf area, it is the United States that explicitly links the crucial factors operating in the security of this composite region.

The only strategically countervailing superpower to the Soviet Union, the United States projects its nuclear deterrent onto the region, linking the nuclear and conventional defence of the southern flank of NATO. At the conventional level of defence, the U.S. Sixth Fleet, together with U.S. Air Force contingents based in Mediterranean countries, furnishes, through its NATO missions, an integrative operational network among territorially disparate members of the Alliance, whose major security links with the United States are primarily bilateral in nature.

As the major producer of advanced military technology

in the West, and its principal exporter to these regions, the United States also creates the most crucial interrelation between technology and security in the entire area. This does not lend itself to battlefield integration, but it does create some possibilities for interoperability. France, Britain and Italy create sources of weapons for local states not formally linked with the West also, but it is the U.S. that is the major alternative to the Soviet Union in technology transfer.

Most important, the United States as the principal mediator in the Arab-Israeli conflict, and the protagonist Western power in the Gulf region, represents the Western nation most capable of providing the political cement that can ultimately bind Western interests in the Mediterranean, the Middle East and the Gulf. To be sure, Europe can and does play an important complementary role. It is unlikely, however, that Europe can supplant the United States as a political broker in intra-regional conflicts.

The global definition of U.S. interests combines with American membership in NATO and the nature of its role in the Gulf to make the United States the paramount link between European security and the Western defence of the Gulf. Conjointly, the Alliance's Mediterranean flank is inherently the operational bridge between the military security of Western Europe and the defence of the Gulf states against possible attacks by Soviet forces or in coping with insurgencies that might create Soviet client states or satellites in that region.

For international security in the 1980s, therefore, the geopolitics of the Mediterranean and adjacent regions will most critically intersect the conflictual East-West political and military interactions and North-South economic and political relations.

THE DEFENCE OF THE GULF AND MEDITERRANEAN SECURITY

There are several levels of interconnections between the defence of the Gulf and security in the Mediterranean region. Most of them cannot be divorced from the security of Europe as well.

In fact, the shift of focus in the American-Soviet rivalry in the various areas of the Third World to the Gulf,

produced by the factors just indicated, has forged strong
and unavoidable links between Mediterranean security and the
defence of Western Europe that go well beyond the formal and
requisite operations in NATO's southern flank. For in the
practical referents of the geopolitics of the 1980s the
defence of the energy resources of the Middle East has
become crucially relevant to the defensive mission of the
Atlantic Alliance.

Because the Alliance was originally designed to defend
solely the areas of the North Atlantic, and because of the
long-standing divergencies between American and West European policies towards the Arab-Israeli conflicts, the European members of the Alliance have strongly resisted, in the
past, suggestions that NATO's strategic missions should
include the safeguarding of oil resources in the Gulf.

European governments, particularly in Mediterranean
countries, are likely to continue to resist an official
extension of NATO's defence perimeter, which would involve
the use of their military forces, if for no less a reason
than the strength of domestic political opposition to the
extension of the Alliance's mandate. But the realities of
the shifting East-West military balance and the inescapable
and critical need for energy faced by the Western industrial
countries will have to eventually reshape European defence
policies to conform to the new geopolitical realities.

Until recently, except for France and Britain, the
approach of all European members of the Atlantic Alliance
towards the defence of the Gulf region had been exclusively
political. More often than not, European states have also
pursued domestic policies in regard to the strategically
important oil resources of that region that have been unilateral, conflicting inter se, and competitive with those of
the United States. The latter themselves being competitive
with those of Europan allies, at the commercial level.

It is a truism of the nuclear era that deterrence of
East-West military conflict is the only rational policy
option for the United States and the Soviet Union and their
respective allies. Some have argued that the extent and
nature of a potential threat to the security of Europe has
been unduly emphasized. Would they maintain their position
after the invasion of Afghanistan, the failure of SALT II
ratification, and the recent developments in the
Euro-strategic and the Soviet- American strategic balance;

with their attending political agitation and the dilemmas raised in U.S.-West European relations?

Others have maintained that the loss of U.S. strategic superiority and the shifts in the Eurostrategic balance, adverse to the West, can be compensated by changes in deterrence doctrine that make possible the conduct of limited and even protracted nuclear exchanges. Indeed, among nuclear strategists the dominant view is that escalation control can be maintained in limited nuclear war. On technological as well as political grounds, how confident can such predictions be? (2)

An examination of the disappearing boundary between the central superpower deterrent system - wrought by evolving technology, shifts in East-West military balances, and redefinitions of superpower strategic doctrines - will be useful in answering questions that arise in the assessment of the significance of the Mediterranean in European security, and of the linkage between the defence of the Gulf and security in the Mediterranean. (3)

The historic relationship between the Mediterranean and Europe gives politics and security in the Mediterranean particular significance for European security. In turn, Mediterranean security, because of the membership of France, Italy, Turkey, Greece, Spain and Portugal in the Atlantic Alliance, cannot be assessed without direct reference to the state of East-West political and military relations in Europe. The changes in military technology and in the U.S.-Soviet nuclear balance and in the NATO-Warsaw Pact military equilibrium during the 1970s have intensified the intrinsic relationship between Mediterranean and European security.

Europe's importance for international and Mediterranean security is explainable not simply by Europe's economic and political weight in the world. It is also the consequence of history, which once made it the central system of international politics. Even after it lost its primacy in world politics, during the thirty-eight years of the nuclear age Europe has been directly linked politically and militarily with the American-Soviet strategic balance, becoming the ultimate pivot for the global rivalry of the superpowers. And it is likely to remain so because of the Soviet Union's location as an Asiatic and European power. Therefore, the fusion of East-West and North-South conflictual factors in the Gulf and the Middle East will not displace

Europe as a crucial and direct link between the central U.S.-Soviet system of deterrence and the Mediterranean regional conflict systems. It is possible, perhaps increasingly probable, that regional conflict in the Gulf might escalate into nuclear conflict. But it is practically inconceivable that an East-West military conflict in Europe would not definitely raise a very severe risk of nuclear war on a global scale.

Therefore, the speed and the character of the changes occurring in the Mediterranean and its regional sub-systems, like the Gulf, would be less grave if they had not been accompanied by adverse changes in the European nuclear and conventional balances - themselves directly related to the technological and structural changes in the strategic relationship between the United States and the Soviet Union. The United States has lost its escalation dominance in Europe and it certainly does not possess it in the region of the Gulf.

Until the 1970s, the military asymmetries in the conventional sphere that have afflicted Western Europe since World War II in regard to the Soviet bloc had been compensated for by the nuclear superiority of the United States and by America's economic strength. These compensatory mechanisms have been eroded and have become part of the problems of the 1980s for European security, particularly in the area of NATO's southern flank, where they importantly influenced the calculations of front line member states like Turkey, by making it possible for them to have armaments that could match, technologically, if not in manpower, Soviet weapon systems.

The dovetailing of these developments in the changing technology of war and deterrence and the momentous political changes that have taken place regarding Iran and Afghanistan have significantly altered the security configurations of the Mediterranean. Quite apart from the impact of changes in the East-West strategic balance, developments in military technology have been reshaping the geopolitical situation of the region.

DEVELOPMENTS IN MILITARY TECHNOLOGY AND THE CHANGING CHARACTER OF SECURITY IN THE MEDITERRANEAN

Changing aircraft and missile technology is shrinking

the Mediterranean - whose North-South axis is already quite short - to the point where land-based systems may totally dominate the sea combat environment. The Soviet Backfire and SS-20 ballistic missiles, stationed in the southern military district of the Soviet Union, can cover the Gulf and Middle East and the whole Mediterranean region; while Western aircraft, sea-based and land-based, can reach them even from the Western quadrant of the Mediterranean. (4) Thus, technology has expanded the range of conventional, and tactical nuclear, regional forces to the point where during acute crises - involving the inferred or actual participation of the U.S. and the USSR - the strategic space can become nearly indistinguishable from the regional one, further weakening the distinction between nuclear and non-nuclear, regional and global.

In terms of nuclear war, there are no genuine strategic military targets in the Mediterranean or nuclear weapon systems strategically significant for the United States. Although there are Western military forces with nuclear capability stationed in several Mediterranean countries and Soviet nuclear systems in the Soviet Union and with the Eskadra that can target the territory of the Mediterranean countries, none of these forces, including those on French aircraft carriers, have strategic missions assigned to them. No American, Soviet, British, or French nuclear strategic launchpoints exist in the Mediterranean. U.S. Sixth Fleet aircraft could reach targets on Soviet territory but they would have to penetrate Soviet air defence and the Sixth could not avoid a Soviet pre-emption. The eventual developments of nuclear-armed U.S. cruise missiles in Sicily will qualify this statement regarding strategic war. At the same time, such a deployment will strengthen the supposition that the strategic space is becoming nearly indistinguishable from the regional Mediterranean one.

This lack of strategic launchpoints does not mean that if an escalating conflict involving American and Soviet forces in this region breached the nuclear threshold - by miscalculation or intent - in the years immediately ahead that nuclear battles could or would not occur. It simply means that there are no strategic territories there in terms of a nuclear central war between the U.S. and the USSR. In regard to the security of the Gulf the United States has officially conditioned the nuclear threshold. Under the

"Carter Doctrine" the United States implied that it might have to resort to the use of tactical nuclears in order to make up for the American conventional military inferiority in the region if the Soviet Union attempted to expand its control beyond Afghanistan, (5) to prevent any major territorial faits accomplis, deemed irreversible. Under the Reagan Administration there have been no references to this eventuality, but neither have U.S. intentions been clarified.

On the other hand, in political terms, certain Mediterranean countries, because of their geographic location in case of conventional military conflict, or their political importance, are strategic. This can be said in the sense that either by being members of NATO or being on the Western side these countries contribute to the deterrence of East-West conventional conflict or that a shift in their political affiliation to the Soviet side would constitute a major political defeat for the West, and weaken Western security. These elements certainly combine in Turkey, Greece, Italy, Portugal, Spain, Israel, Egypt, Yugoslavia, and possibly Morocco.

It is in the realm of conventional armaments that technological changes most pregnant with political consequences have occurred. For example, because of increased aircraft ranges and developments in air-to-surface weapons, no floating aircraft carrier can be a match for the unsinkable platforms represented by the Mediterranean islands, and by the strategically placed Italian peninsula. The current trends in non-nuclear war technologies seem bent on favouring MacKinder over Mahan. Who controls the land, controls the sea. In the Mediterranean and Gulf regions, land-based air power is becoming increasingly threatening to surface warships, increasing the vulnerability of naval squadrons in limited conflict. This will affect their political utility in the management of Mediterranean crises.

Because naval task forces figure prominently in the potential conflicts of these regions, it is pertinent to underscore that the most telling effects of these technological changes have been on surface ships, including aircraft carriers. These are the kinds of forces that would come into play at the outset of a local crisis. The carrier task forces of the U.S. Sixth Fleet have provided a major military tool for crisis manangement in the Mediterranean, and have had the mission of providing U.S. air support for the defence of Italy's frontiers with Austria and Yugoslavia, and for Greek and Turkish air defences. This air support has also provided a backdrop for the defence of Yugoslavia. The increased technological vulnerabilities of such forces have already eroded confidence in the U.S. political commitment to the defence of the Mediterranean. (6) These technological developments have also created a more precarious situation for Soviet ships in the Mediterranean which would have to rely heavily on air cover provided from aircraft based on Soviet territory. (7) Military requirements would also obscure the boundary between the central U.S.-Soviet deterrent system and regional conflicts and would threaten also the distinction between NATO and non-NATO contingencies.

Both the Sixth Fleet and the Soviet Eskadra have had, for quite some time, missions independent of their respective roles in European security. The fact that long-range aircraft equipped with modern air-to-surface missiles, and backed by nuclear-powered attack submarines, currently pose the prime threat to these naval task forces, combines with their alternating redeployment from the Mediterranean to the Gulf to form of these two areas a single conflict system. The operational requirements that the changing relationship between naval and air power is bringing about can have important political repercussions.

The separation of NATO and non-NATO military contingencies has been a major political issue in the renegotiations of U.S. base rights in Turkey and Greece. It has also been an important focus for political pressures that have been brought to bear by the opposition on government policies regarding Italy's role in NATO. With similar rationales the Socialists and Communists, when the former were in the opposition, opposed Spain's entry into NATO on the grounds that Spain would lose its foreign policy independence towards the Third World if it joined the Alliance. Changing military technology and the elevation of the politics of oil to strategic meaning have been creating conditions that could eventually obliterate the distinction beteen NATO and non-NATO military contingencies in the Mediterranean, regardless of the diplomatic positions taken by Mediterranean countries actually, or eventually, hosting U.S. forces and facilities. While technological changes in conventional military systems have tended to diminish the strategic significance of the Mediterranean in terms of U.S.-Soviet nuclear

deterrence, incorporating the Mediterranean into the Eurostrategic deterrent space, they have made the Mediterranean
more crucial to East-West confrontations in the Middle East
and the Gulf. In terms of conventional warfare this has led
to the interlacing, operationally, of the three regions
through the military requirements of the United States as
the principal Western military power in a Gulf contingency.

At the military level the potential for escalation
from even very limited, local political strife requires the
operational readiness of an infrastructure that embraces the
extended Mediterranean region as a whole. Political decisions by Mediterranean states regarding U.S. military operations and the modalities of U.S.-Soviet confrontational behaviour will inevitably come into play, decisively affecting
the security of the Mediterranean region.

NATO ASPECTS OF GULF CONTINGENCIES AND SUPERPOWER RULES OF ENGAGEMENT IN THE MEDITERRANEAN

For the governments of NATO southern flank countries,
decisions involving non-NATO contingencies have been determined, so far, primarily by considerations focussed more on
isolating Mediterranean countries from U.S. military and
diplomatic initiatives in Middle East conflicts than from
considerations of crisis management and local conflict outcomes. The decision of France and Italy to join the U.S. in
sending troops to Lebanon to help resolve the impasse over
the evacuation of the PLO from Beirut could be a harbinger
of change.

It remains a moot question, nevertheless, whether
these policy attitudes would prevail in a Gulf crisis or
conflict involving the United States. There are some indications that do suggest a possible change in the approach of
the European allies to the United States in regard to Gulf
contingencies. For the first time in the history of NATO, a
security interest outside the traditional defence perimeter
of the Alliance has been recognised. In May 1980, the
Alliance agreed officially to a plan enabling the United
States to divert U.S. forces, assigned to NATO, to deal with
emergencies in the Gulf. The following November, the Reagan
Administration transferred the command of the Rapid Deployment Force from the United States to European Command
(EURCOM) in Heidelberg. EURCOM is under the jurisdiction of

the Supreme Allied Commander, or SACEUR. (8)

Thus SACEUR has now become responsible not only for the defence of Western Europe, but also NATO's vital interests outside Europe. Deductively, it is now clear that the U.S. Rapid Deployment Force would expect to rely on NATO facilities, and the redeployment of supplies and U.S. forces from NATO to conflicts outside the European theatre. For practical purposes, this means that NATO's southern flank has been extended from Turkey to the Gulf. (9)

Yet, in the past, the Alliance has known no more divisive issue than the use of its installations by U.S. forces in crises outside the formal boundaries of NATO; connected with threats not directly posed to European security by Warsaw Pact military actions. European governments are likely to continue to apply severe constraints on the U.S. use of facilities in their countries in cases of direct Arab-Israeli conflicts. Yet, are potential conflicts in the Gulf likely to be so neat in their political and diplomatic aspects, short of outright Soviet invasion?

The crucible of the relationship between European security and conflicts in the Gulf may well be forged in the Mediterranean, and be directly linked to the use of air and sea power. The crucial factor will be the participation of Soviet naval forces and Soviet land-based air power.

An escalation from a conflict in the Gulf to strategic, nuclear superpower confrontation so rapid that the conventional military assets of NATO in the theatre, including tactical land-based air power, would become largely irrelevant to the naval battle in terms of its original missions and its contemplated outcomes, could only result from misperceptions or miscalculations, not a deliberate policy choice. If, however, because of the consequences of strategic equivalence and related military and political considerations, escalation, if it occurs, is limited initially to the NATO southern flank level, the role of allied and U.S. land-based air power becomes the most crucial and dominant aspect of the engagement.

The increasing vulnerabilities to air attack of U.S., allied and Soviet navies in the Mediterranean and the Gulf could bring political disaster, in an area where North-South military conflicts are in the offing like the Sahara conflict and the tensions between Libya and Malta. Advanced aircraft are widespread in the countries of the southern shores of the Mediterranean. (10) The day may not be far off when land-launched cruise missiles may also be found among the politically shifting states of the Mediterranean. This outcome would blur even further the distinction between the central system of deterrence and local conflicts in an area where the East-West line of demarcation is fluid and conflictual situations, internal and between states, are on the rise.

The repercussions of Iranian Muslim fundamentalism, the Israeli intransigence on the Palestinian issue, exemplified by Israeli's invasion of Lebanon, and the Shatt-al-Arab war will merge even further the politics and the security of the Mediterranean with that of the Gulf region. Thus, the confluence of political, economic, and security interests, as an expression of the weakening of the boundary between internal and external affairs can become directly connected with the impact of changing technology on the deterrent, crisis management, and warfighting uses of naval and air power in the Mediterranean basin.

On the other hand, the control of most of the riparian territory by formal, or tacit, allies of the United States creates a major potential air threat to the Soviet Eskadra which inhibits Soviet incentives to engage directly American naval and air power in the Mediterranean. In the Gulf, the situation is quite different. The calculations and the requirements of Soviet policy in the Mediterranean and the Gulf are broader than their military aspects. Past Soviet behaviour does suggest, however, that Soviet decision-makers do take seriously the structure of the military situations in the Mediterranean.

The pattern of Soviet crisis behaviour in Mediterranean and Middle East crises shows a clear disinclination to commit military forces to that area. In each major postwar Mideast crisis, the Soviet Union seemingly delayed threats of military intervention until a resolution of the conflict was already fairly clearly in sight. (11) Soviet behaviour in regard to the 1982 conflict in Lebanon is consistent with this assessment.

Although this approach has been a viable way to avoid confrontation and possible escalation to East-West war in the Mediterranean and the Middle East, it does not follow that the United States or the Soviet Union will give up attempts to limit its adversary's efforts to expand political influence and control or desist from the goal of eliminating its rival from the region.

Soviet and American inhibitions against engaging each other's military forces in the Mediterranean region and the Gulf because of the risks raised by superpower military conflict could survive, nevertheless, the blurring of the distinction between the central and regional conflict systems and the weakening of the boundary between internal and international political systems.

Soviet behaviour in past Middle East and Mediterranean crises is, after all, not an isolated phenomenon. It stems from recognisable anxieties and doubts regarding the ability to control the local conflict situation so that it does not escalate to strategic confrontation and war. The difficulty of calculating U.S. crisis behaviour is also a factor, because U.S. behaviour during crises has lacked steadfastness and predictability, over time.

The behaviour of U.S. decision-makers during crises must surely be baffling to Soviet leaders all the more because it has been in the United States, particularly, that refined theories on the calculated use of strategic power and coercive diplomacy have been developed. Some of these doctrines have been incorporated in American strategic policy; like the limited war concepts of Presidential Directive No. 59, and presumably the National Security Decision Document 13. (12)

Similarly, refined theories about the manipulative use of strategic power, like compellence, graduated escalation, and selective nuclear options, have not been developed in the Soviet Union. Soviet reluctance to use military force beyond the immediate periphery of Soviet hegemony are attested by the Soviet invasions of Hungary and Czechoslovakia, in 1956 and 1968, the Sino-Soviet border clashes of 1969, and the more recent Soviet invasion of Afghanistan. (13) The single exception, the Cuban missile crisis, has been recognised by the Soviets as aberrant and dangerous, resulting in the fall of Kruschev from power.

At the same time, it is important to remember that neither the United States nor the Soviet Union can tolerate the decisive defeat of a major ally, or client state, by external forces that are a proxy of the adversary superpower. Such defeat would significantly affect its own regional position and its credibility as a patron. This explains, in part, why the simultaneaous involvement of the superpowers in the Middle East has added a significant dimension to the regional problem rather than simply a means of resolving it.

These facts, coupled with the fluidity of the situation, also underscore why the danger that local war might escalate into superpower, and East-West conflict, remains very real. The concept and the operation of mutual deterrence that has prevailed in Europe is difficult if not practically impossible to operationalize beyond the southern flank of NATO. Superpower tacit confrontational rules in third areas are all that is available for dampening the prospects of escalation to superpower conflict in the southern reaches of the Mediterranean and the Gulf.

These rules have as their primary objective the avoidance of escalation from local conflict to general nuclear war between the American and Soviet superpowers. The avoidance of direct military engagements between Soviet and American forces in local conflicts operationalizes this maxim. As a consequence, it has been possible, so far, to keep non-European concerns from becoming central to the American-Soviet rivalry in ways that would threaten military conflicts in Europe.

Another tenet that has operated within these rules, since the 1962 Cuban missile confrontation, has been the mutual recognition that the line of superpower competition must be drawn short of the vital U.S. and Soviet interests defined by their respective spheres of political control, with the prerogative of each superpower to protect its interests. (14) But where does one draw this line in the Southern and Eastern Mediterranean and the adjacent Gulf?

The most forthright declaration defining United States interests in this extended region is that of President Carter. It continues to be the basis for U.S. policy. On January 23, 1980, in his State of the Union speech Carter declared: "An attempt by any outside force to gain control of the Persian Gulf region will be regarded as an assault on the vital interests of the United States of America and such an assault will be repelled by any means necessary, including military force". (15)

However, neither the geographic nor the political boundaries of the Gulf and the Middle East are precise enough to warrant the assumption of a clear demarcation line between East and West, as is the case between NATO and the Warsaw Pact.

That line has tended to become obscured during conflicts in the Middle East. For example, during the 1973
Arab-Israeli conflict, Turkey permitted Soviet military over-
flights and Soviet naval transit through the Dardanelles to
resupply Egypt and Syria. At the same time, Turkey refused
the U.S. permission to use facilities on Turkish soil, to re-
fuel or reconnoitre during the U.S. airlift to Israel. The
Turkish leadership's sensitivities, regarding their independence of action, were focussed more on their proximity to
the Soviet Union and their reluctance to provoke a Soviet
response against Turkey than on concerns regarding their
standing with Arab states. The main Turkish preoccupation in
an eventual Gulf crisis would be whether it involved an
actual or potential Soviet client state. (16)

The prudence displayed by Soviet leaders regarding
military initiatives of questionable strategic value is likely to continue to assert itself in situations that could
lead to strategic engagement with the United States. (17) On
the other hand, visible weaknesses in an adversary's military capabilities have historically created incentives for
quick military initiatives. All the more so if the political
payoffs were significant, and timely retribution difficult
to effectuate.

Soviet leaders are not uniquely bound by previous
patterns of behaviour, notwithstanding their belief in the
inevitability of communism and their commitment to Leninist
tactics which bridle radical short-term solutions that might
undermine long-range strategic goals.

The evident weaknesses in U.S. military capabilities
in the Gulf area and, for the time being, in the reserve
situation as well - coupled with the geographical advantages
of the Soviet Union - could generate incentives that would
erode Soviet caution. This outcome would be facilitated if
political opportunities arose which were sufficiently tanta-
lizing or threatening to perceived Soviet national interests. The experience the Soviets have been having in Afghanistan may be most pertinent in this respect. It may be too
soon and too difficult, with the evidence at hand, to assess
how critical this factor might be, however, in Soviet decisions during future crises in the Gulf region. Equally
critical would be the state of the political relationships
that prevailed, at the time, between the United States and
its NATO allies, and between the United States and countries

in the region of the Gulf.

The very political factors that operate to create binding commitments between the United States and regional powers in the Mediterranean, the Middle East and the Gulf threatened by Soviet hegemonial tendencies - be they direct expansion by military means or indirect control gained through supporting the overthrow of the governments in these countries - also produce direct links between local turmoil and superpower confrontation by coalescing internal and international politics into a single process.

The southern Mediterranean and the Gulf are regions that exhibit precarious inter-state dynamics either because national identities are at least in part uncertain or, even more importantly, because the legitimacy of governments is often qualified. Nation-building in these regions, with its attendant necessity for political and economic develoment, is threatened not only by ethnic fragmentation, but also by an ideological search for the means to achieve a national cohesion that will legitimize the rule that must guide political, economic, and social development into a modern mold, capable of incorporating also the essence of traditional values.

The most critical links between internal and international politics of relevance for international security are forged in this nexus. For it is there that superpower intervention is politically rationalized in the context of rival ideologies by the leaders of the United States and the Soviet Union, as well as by indigenous ruling elites.

The U.S. intervention in Vietnam, in the 1960s, and the more recent Soviet intervention in Afghanistan are clear examples of the dual nature of the relations between superpowers and local states. From one angle, a clear and firm defensive relationship between the patron and client states acts as a deterrent to the adversary superpower. From another, if internal political circumstances obscure expectations about that relationship, miscalculations with grave consequences for peace are possible, especially if the military capacity of the status quo superpower is inadequate or severely hampered politically.

It has been pointed out most cogently that the success of the security-related commitments undertaken by the United States to safeguard the principal oil supply for the Western and industrial world cannot depend on U.S. will alone, nor on primary reliance on military instruments alone. "In an area as fragmented, unstable, politicized, ... as well as vulnerable to, external intrusions as is the Middle East, a defence policy that relies in the main on a military instrumentality to create collective security is bound to prove inadequate, unless the political foundations on which a multinational effort must rest are solid." (18) Otherwise, the regional use of American military forces can be frustrated in its objectives "in the moment of need by the unwillingness or inability of local governments to do their share." (19)

In effect, deterrence of the possible use of military means to achieve Soviet political aspirations in the oil-rich Gulf region, and the concomitant risks of escalation to superpower confrontation or war depend critically on the military and political viability of U.S. approaches to the situation that prevails in these and in adjoining regions.

Consequently, it is a matter for concern that the juncture effected by the U.S. between the defence of the Gulf and the Atlantic Alliance, particularly with NATO in the Mediterranean, derives from weaknesses in U.S. military posture and the restrictive assumptions of U.S. diplomacy.

To these weaknesses must be added the dysfunctional impact of the disagreements between the United States and its West European allies on the diagnosis of the security threat to the Gulf and the appropriate means to resolve it.

The concrete military elements that make the United States the link between NATO and Gulf security derive, in part, from U.S. handicaps not easily compensated for by its allies. Setting aside, for the moment, the ways in which U.S. NATO allies could assist the United States as the guarantor for the defence of the Gulf, the main U.S. shortcoming is definable in terms of the availability of appropriate military forces.

Although the establishment of sufficiently equipped facilities in the Mediterranean and the Gulf for use by American military forces is a critical aspect in the defence of the Gulf, an equally major shortcoming is the deficiency in American military capabilities. The decision to create a U.S. Rapid Deployment Force from existing military units, mostly earmarked for NATO and the Far East, has widened the gap between U.S. commitments abroad and the military resources needed to fulfil them. By diminishing the U.S. military capacity in NATO, should threats to the Gulf region require it, the security of Western Europe would be degraded precisely at a time of crisis when a strong deterrent against the temptation for possible Soviet pressures would be needed.

U.S. military planning notwithstanding, current American force levels make it virtually impossible to deal effectively with a significant military challenge in more than one area. (20) Thus, an intractable problem now faces the West. How is the U.S. to implement a simultaneous force buildup in NATO and the Gulf if prudence requires it? U.S. force levels are not sufficient to repel a Soviet assault in the Gulf without jeopardizing the U.S. commitment to the conventional defence of NATO. (21) The war in Vietnam was waged, to a considerable degree, by U.S. forces earmarked for European contingencies. In 1980, the deployment of U.S. carrier task forces in the Arabian Sea was made possible by withdrawing U.S. carriers on station along NATO's southern flank and in the Pacific. (22) Inversely, the 1982-1983 U.S. naval presence off embattled Lebanon was augmented by U.S. naval forces from the Indian Ocean and the Atlantic.

The strategic risks resulting from reliance on U.S. forces committed to both Gulf and NATO contingencies could be serious in the event of a confrontation between the United States and the Soviet Union. The Soviet Union benefits from shorter and interior logistical lines of communication, larger military forces and contiguous proximity to the Gulf, as well as Europe. Soviet pressures in one area could divert deployable U.S. forces from the real location of Soviet intent. (23)

It should not be surprising, in view of these factors, that of the problematic relationships which connect NATO's presence in the Mediterranean with the role of the United States in the Middle East and the Gulf, security considerations have been central in American policies and crucial in the relations of local states with the American superpower. Politically, the United States remains the dominant external power in the Mediterranean and the adjacent regions.

The United States plays the crucial role for Western security in the Mediterranean. It is the only country that has been able to generate sufficient countervailing military power to balance the projection of Soviet military power into the region. Moreover, until recently the U.S. has been the Western nation most capable of discharging the role of conflict manager. (24)

For the forseeable future, it remains the only major external power with the political capacity, through its bilateral relations within and outside NATO, to bring coherence to the defence of the area. No European nation seems capable of substituting for the United States in this role.

U.S. NATIONAL INTERESTS LINKING THE MEDITERRANEAN TO THE GULF AS A FUNCTION OF EAST-WEST FACTORS

It has been argued that American foreign policy has been torn between universalist policies with reliance on narrow security interests served by military power and "abstention, value judgments and world order." (25)

Insofar as U.S. national interests in the Mediterranean and the Gulf are concerned, geopolitical considerations have by far and large dominated. In fact, it can be maintained that the only real tensions in U.S. policies towards these areas have occurred between the global thrust of American foreign policy and the inevitably local concerns of regional states. Otherwise said, between the requirements of the central, superpower system of deterrence and the difficulties attending regional security, in areas where local enemies are more immediately threatening to regional states than is the military power of the Soviet Union.

These tensions have persisted and lately intensified because the military reach of the Soviet Union has grown in capability. The intimate relationship between the exploitation of regional conflicts for Soviet political purposes and the East-West military balance persists. It will continue to confront the United States with difficult choices about direct involvement in regional crises and conflicts. For it is not readily apparent when U.S. involvement will strengthen the East-West equilibrium in these areas or when it will sap U.S. military capacities, and therefore the diplomatic credibility of the United States.

For all that, basically the United States had viewed regional military balances as critical even before the advent of strategic parity summoned the fear that the Soviet Union would exploit this condition for probing initiatives in the Third World, where they could become critical for the overall East-West balance of power. However, shifts in regional balance in third areas are most vulnerable not to superpower manipulation but to internal political changes. The latter are not amenable even to prescriptive superpower military interventions, because they are political outcomes of more complex social changes, not mere changes in ideology or political regimes.

These the superpowers can exploit but not generate. The fall of the Shah in Iran and the expulsion of the Soviets from Egypt, as well as the internal political changes in Afghanistan which confronted the Soviets with the dilemmas and the opportunities that led them to invade, are not primarily explainable in terms of failures or successes of U.S. foreign policy.

The same can be said for the fundamental regime changes that have occurred in Portugal and Spain. On the other hand, a strong argument can be made that Soviet interference in these cases was kept at bay by their location in the U.S. security sphere. Similarly, Egypt's situation in an essentially American sphere of influence surely inhibited an aggressive response to Sadat's political volte-face. Even Khomeini's revolution, at least up to the hostage crisis, must have benefited by Soviet calculations regarding the linkages between Iranian national defence and U.S. security interests. On the other hand, Afghanistan's historical non-alignment, coupled with a revolutionary regime change, did not present the Soviets with such clear-cut inhibitions.

At the same time, American national interests in the Mediterranean and the Gulf region cannot be understood if they are divorced from the centrality of East-West considerations in U.S foreign policy. That is, the basic referent for American politics towards these regions has been the containment of the Soviet Union's hegemonial strivings. Moreover, although U.S. policies have certainly included awareness of the importance of the political and economic aspects of regional problems, the emphasis during most of the postwar period in U.S. policies towards the Mediterranean and the Gulf has been on security, (26) while local political constraints on American military operations have been essentially political. They have derived from the internal politics of Mediterranean and Mideast countries, and have been applied during conflicts between regional states.

Throughout the postwar period, Washington has viewed conflicts like the Arab-Israeli wars and the Greek-Turkish conflicts in Cyprus as creating opportunites for Soviet diplomacy to expand the political influence and the military presence of the Soviet Union in the Mediterranean and the Middle East. This definition of U.S. national interests in these regions in markedly security terms has had a basic consistency, easily brought out through an historical perspective. Notwithstanding the deviations, exemplified by the dysfunctional U.S arms embargo against Turkey during the late 1970s and the emphasis of the initial policies of the Carter Administration, American governments have shown a persistent preference for geopolitics and global, or East-West, approaches. The historical record is both clear and instructive.

The American military presence in the Mediterranean, and American diplomatic interest in the Middle East, coincided with the transformation of the United States into a global military power with a permanent internationalist foreign policy. It may be argued, in fact, that the first act of U.S. policy that could be qualified as Mediterranean was the enunciation of the "Truman Doctrine". In 1947, U.S. economic and military aid to Greece, torn by a civil war initiated by pressure at the Dardanelles and Turkey's eastern frontiers, became the first concrete acts of assistance to local states, launching U.S. containment. Britain having given notice to the United States that it would be unable to continue economic and military support for Greece and Turkey, the U.S. government saw this as a situation where the concept of containment was reasonable.

As for the Gulf region, in 1946, following the end of hostilities between the Axis and Allied powers, the Soviets refused Allied demands to withdraw Soviet forces from northern Iran. They rationalized their position by the establishment of the "autonomous" republics of Azerbaijan and Kurdestan. The American government induced the Soviet leadership to withdraw by making clear to them that the United States was "too strong to be beaten and too determined to be 'frightened." (27) U.S. geopolitical interests, because of the Soviet Union's proximity to the Gulf and the Middle East, dictated concern about Soviet moves even without national consideration connected with oil. Nevertheless, U.S. preoccupation with Soviet conduct in the Gulf region did not lack awareness of the security implications of oil, even then.

The Truman Doctrine's political and military rationales for aid to Greece and Turkey were leavened by this turn of events. Mutual U.S. security treaties were concluded through a bilateral approach that was to define the future orientation in U.S. relations with the countries of the Mediterranean, including those joining NATO, and in the Middle East. The U.S. bilateral accords for naval, air, and intelligence installations, signed during the 1950s with Italy, Spain, Morocco, Portugal, Iran, and Libya, need no elaboration. Clearly, these were cast in an explicitly East-West mold.

If defined from the perspective of the various American presidents since Truman, the official objectives of U.S. policies towards the Mediterranean and adjoining regions remain essentially those the United States set forth when it first entered Mediterranean and Middle Eastern politics:

- To maintain a balance of power with the Soviet Union.
- To help defend Greece, Turkey, Italy, Spain and Portugal against direct and indirect military and political pressures by the Soviet Union.
- To guarantee the survival of Israel (and after the expulsion of the Soviets from Egypt, also Egypt's independence).
- To help assure the flow of vital oil to Western Europe, and the West.
- To keep the Soviet Union and Soviet influence out of the Middle East and North Africa.
- To promote regional stability in the Mediterranean and the Middle East.
- To maintain the political cohesion of the southern flank of NATO. (28)

A threat to U.S. interests in the Mediterranean does not have to come directly from the Soviet Union. Regional states can create, on their own, political instabilities and crises affecting Mediterranean security. These include revolutionary and interstate conflicts that can adversely affect U.S. national interests by creating opportunities for Soviet political influence and military penetration. U.S. tolerance for the export of revolution or the use of confrontational tactics and outright military combat by regional antagonists is necessarily limited in a region where a U.S.-Soviet military collision is always possible and where geography and politics link with possible threats to NATO and European security to the north and conflictual oil politics and

Soviet expansion to the South.

Thus, the original premises of the Truman Doctrine and the first containment policies linking indigenous communist political forces and Western security in these areas were never abandoned by successive American administrations. Only once, at the outset of the Carter Administration, but fleetingly, in regard to Eurocommunism, did the U.S. approach deviate. Even then, however, the change was in tactics not strategy. In any case, the main U.S. concern about Eurocommunist parties was always on security (29) aspects. The Carter approach was predicated on a permanent division of influence along East-West lines in Europe and the Mediterranean. As intractable as the issue of Eurocommunism has been for U.S. foreign policy, its resolution, in American foreign policy, has demonstrated the strength of the traditionalist outlook in U.S. policy towards the Mediterranean; firmly planted in East-West considerations.

The centrality in U.S. Mediterranean and Gulf policies of the rivalry with the Soviet Union, with its emphasis on the political uses of military force, has inevitably cast the United States in the role of ultimate guarantor against Soviet military threats to the security of the Mediterranean countries of Europe and Turkey, members of NATO, and Israel and of West-leaning Arab countries. Political turbulence in these regions has also forced the United States to become the crisis manager and the mediator in regional conflicts - lest the security of the area be undermined by providing the Soviet Union with diplomatic and subversive opportunities.

These U.S. roles have not merely operationalized in forceful fashion American policy objectives in these conflictual and politically unstable regions, they have been - and will continue to provide - tests for the viability and success of U.S. national interests. An admittedly terse summary of the past record of the United States in the role of security guarantor and conflict manager and mediator suggests the United States had successfully managed these roles, until the mid-1970s.

The 1974 Greek-Turkish conflict in Cyprus, and its aftermath, have cast a long shadow over American capability to mediate Mediterranean conflicts among allies. The radical changes brought by the fall of the Shah in Iran, the Soviet invasion of Afghanistan, and the Shatt-al-Arab war have also seriously questioned U.S. capabilities both as security guarantor and conflict manager in the Gulf. Only in the Middle East does U.S. ability to perform as a conflict manager and mediator remain purposeful. Nevertheless, it remains a moot question how the conflict in Lebanon will affect this role in the future. The success or failure of U.S. policy in Lebanon will crucially shape the future effectiveness of the U.S. as the diplomatic broker in the Arab-Israeli conflicts. As for an American security shield against possible Soviet military encroachments in the Eastern Mediterranean, U.S. resolve has not been tested since the 1973 October War.

ISSUES AND PROSPECTS FOR SECURITY

Given the political and military developments that relate Mediterranean security to defence of the Gulf, what can be said about alternative options for the American military presence in the Mediterranean? A question that must also be sensitive to the political and economic factors of the region.

This issue has two related but divisible aspects. One concerns the deterrence of East-West conflict in Europe and of the possible or potential Soviet military expansion into the Gulf region; for example through Baluchistan. The other addresses local conflict resulting from attempts to radically alter the political regimes of Gulf and Middle East countries - like those that have occurred in Iran and Ethiopia - but which would also deny the West an important source of oil while realigning the country politically with the Soviet Union. The key state, in this context, is Saudi Arabia.

The American military presence in the Mediterranean region has been, and continues to be, definable almost exclusively in terms of naval and air power. The U.S. Sixth Fleet is the most visible American military capability there. It represents warships, seabased air power, and marine infantry combined into the most powerful, immediate and flexible U.S. combat contribution to the defence of NATO's southern flank. The planned deployment of U.S. nuclear capable cruise missiles in Sicily underscores the U.S. Sixth's deficiency as a strategic contribution to the deterrence of nuclear war.

In terms of the Mediterranean, the Soviet Union with the Backfire, Fencer, Flogger D has deployed relatively sophisticated aircraft that can carry nuclear weapons deep into NATO territory, and into the Middle East and the Gulf. The mobile MIRVed SS-20 missile adds a Eurostrategic dimension not matched by any U.S. weapon system deployed in these regions. (30)

The inescapable vulnerabilities of the Sixth to Soviet pre-emption in the case of nuclear conflict do not negate, however, its utility to deter conventional war between Warsaw Pact and NATO forces in the Mediterranean, or in fighting one there. An outbreak of conflict directly involving NATO territory should remove any political restrictions on the use of American and allied land-based air power that would assist the Sixth Fleet in protecting itself from air and submarine attacks, while delivering its own attacks on enemy sea and land targets.

Moreover, when compared with the naval and air technology available to the Mediterranean members of NATO, there is little question that American sea and air power provides the only match to Soviet military technology in the region. No cogent political arguments can be made against the continued contribution the Sixth Fleet makes in the deterrence of East-West conventional conflict in the Mediterranean. Nevertheless, prudence might require that because of the increased vulnerability of naval surface ships to land-based aircraft the Sixth Fleet should remain deployed in the Western Mediterranean during the initial phase of a conflict, or until Western forces have achieved command of the airspace over the Eastern Mediterranean.

There remain concerns about the potentiality that direct naval engagements between the Soviet and American Mediterranean fleets, and their supporting sea-based and land-based aircraft, might cause escalation to nuclear fire. Certainly, the more intense technological impact of land-based air has enlarged the geographical operational requirements of the battle zone from the very onset of a battle. The use of airfields in Turkey, Greece, Italy, Spain, Bulgaria and the Soviet Union has become more immediately crucial.

In a clear-cut NATO-Warsaw Pact confrontation, devoid of encumbering local conflicts, the Mediterranean deployment of the U.S. Sixth Fleet does not, of itself, increase the chances for escalation to nuclear conflict - unless one can argue that the permanent presence of Soviet ships there poses neither military nor political threats to NATO. Such an argument would be strictly ideological because it would reject any connection between military force and political purpose.

Provided a prudential command or arms control approach is followed by Soviet and American naval task forces, the presence of the Soviet Eskadra and the U.S. Sixth on NATO and Pact missions do not significantly increase the risks of nuclear war. To the contrary, they provide intermediate conventional options not otherwise available for mutual deterrence. Under such circumstances the crossing of the nuclear threshold, by either side, would be a deliberate political decision that could not be taken, rationally, without full cognisance of the risk of escalating to nuclear war between the Eastern and Western alliances.

The blurring of the boundaries between conventional and nuclear deterrence systems and conjointly between the regional and the central superpower systems of security which developments in military technology, aggravated by the confluence of American and Soviet nuclear warfighting doctrines, are threatening to bring about are not the consequence of Soviet and American naval deployments in the Mediterranean and the Gulf. They cannot be resolved through redeployments there although they do affect directly those deployments.

The changed technology (i.e. range, accuracy and mobility) of Eurostrategic, land-based weapon systems, on both sides, are a more significant menace to the nuclear threshold. These land-based nuclear forces do not have the flexibility of being easily redeployed in and out of the European theatre of operation.

Even more salient, although the Sixth and the Eskadra are capable of both nuclear and conventional missions, they have acquired a predominantly deterrent and especially conventional image. Neither has a credible nuclear first-strike capability except against each other. The U.S. Sixth Fleet has also projected, through the nature of its use over the years, the almost exclusive connotation of being the preferred military instrument for U.S. crisis management in the Mediterranean. The clearly American identity of this task force, and its non-territoriality once at sea, have facilitated a political disengagement from Arab-Israeli crises by NATO allies which a militarily meaningful deployment to the Gulf of the U.S. Rapid Deployment Force might not allow, especially in the incipient phase of any local conflict, because of the force's transit and logistical requirements.

This means that the Sixth is not handicapped by having to provide an explicit linking between the deterrence of nuclear conflict in Europe and the deterrence of nuclear war between the Soviet and American superpowers. Conversely, it does not face the dilemma of raising, by its deployment in regional crisis, the danger of decoupling the security of Europe from that of the United States.

In light of these observations and vis-à-vis the deterrence of nuclear and conventional war no useful purposes would be achieved by withdrawing the U.S. Sixth Fleet from the Mediterranean. I now turn to the examination of the relevance of the Sixth Fleet to the deterrence of possible Soviet expansion into the Gulf area and its interface with the defence of the Gulf area and the security of the Mediterranean.

Reasoning from a purely military perspective, and without consideration of specific factors arising from indigenous conflict considerations in the Gulf area, the U.S. Sixth Fleet could contribute to the deterrence of possible Soviet incentives to pursue expansionist territorial goals in that region or to direct Soviet assistance to secure a rebel regime that had overthrown a local government aligned with the West.

The Sixth's deterrent utility derives from its ability to redeploy quickly at least one carrier task force into the Gulf, via the Suez Canal, and to possibly interdict Soviet airlifts that came within range of its aircraft. These capabilities should dampen Soviet incentives to use military force, especially during the incipient phases of a local conflict. A complication arises for the fact that even more than is the case in the Eastern Mediterranean, the further elements of the Sixth penetrate the Gulf, the more they become vulnerable to attacks from land-based Soviet air power. In the waters of the Gulf itself the Sixth could become vulnerable indeed. Consequently, U.S. naval task forces in the Gulf would diminish their vulnerability, hence increase their deterrence value, only if they could be supported in their own defence by U.S. land-based aircraft. The latter's most convenient basing is likely to be in Saudi Arabia, Egypt, Israel, Turkey, and Somalia.

Whether the characteristics of airfields in these countries would effectively meet with the requirements of suitable aircraft is beyond the scope of this analysis. It is at least arguable, however, that it would be useful to have sufficient prepositioning of material in case of a prolonged engagement. These requirements cannot be isolated from the regional infrastructure needs of the U.S. RDF. In fact, the requirements of the Sixth when operating in the Gulf, and those of the RDF would have to be mutually reinforcing. This is particularly so in the initial phase of a crisis when the Sixth, by being more immediately available, would be a more critical factor in deterrence.

The technical, organizational and political problems besetting the early development of a suitable RDF, coupled with the Sixth's operational ability to disengage itself from NATO missions and a NATO image, should offer U.S. allies in the Mediterranean sufficient opportunities to consider their political positions during crises, and the early phases of a U.S.-Soviet conflict.

Except in the unlikely circumstance that Gulf oil is no longer essential to the European economies and to the security of Western Europe, it would seem that a clear and direct Soviet threat to the oil-producing countries of the Gulf should issue into positive political support for the military actions the United States might have to undertake in the area of the Gulf.

Two qualifications apply, one military the other political. On the military side, the redeployment of the Sixth from the Mediterranean to the Gulf diminishes the deterrent strength of NATO's southern flank at a time when it may be most needed. The Mediterranean members of NATO and France would have to go well beyond the currently planned remodernization of their naval and air forces to effectively replace the military weight of a two-carrier task force Sixth Fleet. On the political side, and directly related to a Soviet-American confrontation shorn of indigenous conflict, is the fact that not only would the Sixth require more critical back-up from land-based air forces but that it would be part of an operation that included the deployment of the RDF to the region. The latter would more directly include NATO South from the beginning. Portugal and probably Spain could hardly escape immediate involvement. Faced by Soviet capabilities which when combined with the terrain features of the area, vastly increase its own military vulnerability, the Sixth becomes another link in the chain that unites the security of Western Europe and the Mediterranean with that of the United States in the Gulf.

But this potential outcome does not result uniquely from the dispatch of the Sixth from the Mediterranean to the Gulf region. Another U.S. naval task force approaching the Gulf from the Indian Ocean, under similar circumstances, would trigger the same coupling mechanism. More crucial are the links with Mediterranean security that flow from unavoidable requirements connected with the deployment of the RDF to the Gulf.

In regard to an overt Soviet conventional military threat against a member of the Atlantic Alliance in Europe and the Mediterranean it is militarily infeasible and politically questionable to separate the fate of Europe from that of the United States. Moreover, provided Western military postures are sufficiently deterrent and the political cohesion of the Western alliance sufficiently firm, the Soviet use of military force against NATO in the Mediterranean is unlikely.

In the area of the Gulf, the situation is more problematic precisely because, vis-à-vis the Soviet Union, countervailing military capacity is insufficient and politically hobbled. The European allies of the U.S. have little to contribute beyond the possible use by U.S. RDF forces of selected infrastructure sites in NATO's southern flank and political solidarity with the United States against Soviet designs.

However, there too, under present circumstances, it seems unlikely that the Soviet Union would undertake military initiatives against Pakistan, Iran, or Saudi Arabia or other oil-producing Arab states. In geo-political terms, nevertheless, the development of a militarily adequate and politically viable U.S. military capability that can respond in time to Soviet initiatives is a sine qua non for regional deterrence of major power conflict.

Notwithstanding the increased threats to stability on the East-West axis to security, it is the changes that have been occurring along the North-South axis that are the more forboding for the security of the Mediterranean region. These changes are beginning to intrude into the factors that operate in East-West and superpower military and political balances in a way that obscures and complicates the calculus of deterrence.

Regional conflicts have all along threatened the stability of East-West security in the Mediterranean. But before
the political use of oil and the advent of revolutionary
Muslim fundamentalism the relations of the United States
with its Western European allies managed to separate Middle
Eastern conflicts from the vital national interests on both
sides of the Atlantic.

As security and economic interdependence have become
more blended, the political cohesion of the Atlantic Alliance has become more problematic. The relationships between
European and American national interests has become more
complex, more contradictory and potentially more competitive
at the intersection of the East-West and North-South axes of
international politics.

The future security of the Mediterranean will be decided more in the crucible of American and European political
and economic relations, as they clash and converge with each
other, in North Africa, the Mideast and the Gulf, than by
the correlation of forces between the United States and the
Soviet Union and their allies - as critical as these are for
the security of Europe.

NOTES

1. Subnational conflict in the Mediterranean is examined by Brian Jenkins in chapter six of this book.

2. Desmond Ball, Can Nuclear War Be Controlled?, Adelphi Paper 161, IISS, London, 1981.

3. An exploration of the significance of the weakening of the boundaries between global and regional systems is in Ciro E. Zoppo, The Security and Politics of the Mediterranean and International Security, unpublished paper presented at the UCLA conference on "The Mediterranean and World Politics", Rome, December 1980.

4. P.M. Dadant, Shrinking International Airspace as a Problem for Future Air Movements, The Rand Corporation, R-2178-AF, January 1978.

5. Remarks of Secretary of Defence Harold Brown in press conference on U.S. defence of the Gulf, 1980.

6. S. Silvestri a M. Cremasco, Il Fianco Sud della NATO, Feltrinelli, Milan, 1980, p. 116ff.

7. F.S. Nyland, Air-Launched Missile Attacks on Ships, The Rand Corporation, P-6264, December 1978.

8. C. Coker and H. Schulte, "A European Option in the Indian Ocean", International Defense Review, January 1982, p. 1.

9. Ibid., p. 3.

10. See chapter seven by Maurizio Cremasco for an extended analysis.

11. Francois Fukuyama, "Nuclear Shadowboxing: Soviet Intervention Threats in the Middle East", Orbis, Fall 1981, pp. 579-605.

12. Los Angeles Times, 25 August, 1982, p. 6, and 26 August 1982, p. 7.

13. Benjamin S. Lambeth, Risk and Uncertainty in Soviet Union Deliberations about War, The Rand Corporation, R-2687-AF, October 1981, p. 9.

14. Abraham S. Becker, The United States and the Soviet Union in the Mediterranean, The Rand Corporation, P-6039, November 1977, p. 21.

15. Weekly compilation of Presidential documents, 28 January 1980, Vol. 16, No. 4, pp. 194, 197.

16. An examination of the impact of changing Turkish-Soviet relations is in: Ciro E. Zoppo, "Turkey in Crisis: Implications for the Atlantic Alliance", in Steven L.

Spiegel, (ed.), The Middle East and the Western Alliance, Allen and Unwin, London, 1982.

17. Lambeth, op. cit., p. 8ff.

18. Paul Jabber, "U.S. Interests and Regional Security in the Middle East", Daedalus, Autumn 1980, pp. 68-69.

19. Ibid.

20. Jeffrey Record, The Rapid Deployment Force and U.S. Military Intervention in the Persian Gulf, Institute for Foreign Policy Ananlysis, Cambridge, Massachusetts, February 1981, pp. 36ff.

21. Ibid., pp. 43ff.

22. U.S. Department of Defence, "Authorization for Appropriations for Fiscal year 1981", Hearing before the Committee on Armed Services, Part 6, p. 3275.

23. Ibid., pp. 745, 785.

24. Pierre Hassner, "American Policy Toward the Soviet Union in the 1980's", in America's Security in the 1980's, Part II, IISS, London, Spring, 1982.

25. Shahram Chubin "The United States and the Third World: Motives, Objectives, Politics", in Third World Conflict and International Security, Part II, IISS, Summer, 1981. This section of analysis is influenced throughout by Chubin's discussion.

26. An expanded discussion of U.S. Mediterranean policies is in Ciro E. Zoppo, "El Mediterraneo en la Politica Exterior de los Estados Unidos", Revista de Estudios Internacionales, Madrid, Vol. 3, No. 1, January-March 1982, pp. 69-95.

27. Cited in John L. Gaddis, Strategies of Containment, Oxford University Press, New York, 1982, p. 22.

28. Committee on Foreign Relations, U.S. Senate, United States Foreign Policy Objectives and Overseas Military Installations, Washington, D.C., April 1979, pp. 44-45.

29. Ciro E. Zoppo, "Foreign and Security Policies of the Italian Communist Party", Ch. 3 in Lawrence L. Whetten, (ed.), New International Communism, Lexington Books, Lexington, 1982, see pp. 51-56. (The current printing indicates Whetten as the sole ostensible author of the book. This mistake is to be rectified to include the true contributing authors of the chapters).

30. T.L. McNaugher and T.M. Parker, Modernizing NATO's Long-Range Nuclear Forces: An Assessment, Rand P-6486, Santa Monica, Ca., October 1980.

Abdul Karim, Tayeh 23
Abu Dhabi 31,136
Abu Musa (Island of) 136-7
Adriatic Sea 2,7,13,36,41,
  47,67,76,106,116-7
Aegean Sea 2,7,13,41,52,58,
  60,66,70,76,112,119,127,
  206,208-9,211,247
Afghanistan 245-7,293,296,
  298,305,308,312,315
Africa(n) 2,15,41,52,97,129,
  161-2,172,199,244,253,257,
  293
Albania(n) 13,53-4,65-7,75,
  82,100,104-5,113,181,192,
  206,214,216,241
Alboran Sea 60,70
Algeria(n) 3,10,15-6,26-7,31,
  33,36,51,68,70,75,100,102,
  120,134,149-51,153-63,165,
  167,169-75,179,181-2,184,
  192,194,196-7,207,212,
  215-8,245,248
  Arzew LNG installations 26
  Habibas well 3,10
  Hassi R'Mel oilfield 26
  Oued Mellah 36
Algerian Army of National Liberation (ALN) 169
Algero-Moroccan war of 1963
  155,197

Amoco (Standard Oil of India-
  na) 10
Aqaba (Gulf of) 129-30,134,
  209
Arab countries 33,126,139,
  166,172,194,210,214,220,
  242-4,255,257,259,307,
  315,321
Arabian Peninsula 138,142
Arab-Israeli war 280,293,
  295-6,303,306,312,316,318
  Six Day War (1967) 182,197
  October War (1973) 199,210,
  316
Aramco 19
Archipelagic State 96-7
Argentina 222
Armenia(n) 185
  Armenian Secret Army for
  the Liberation of Armenia
  (ASALA) 183
Asia, Southeast 26,41,220,293
Assad, Hafaz 201
al Assifa 184-5
Association agreements bet-
  ween the EC and Mediterran-
  ean countries 246-7
Atlantic Alliance - see NATO
Atlantic Ocean 3,10,47,49,53,
  57-8,76,97,105,114,150,170,
  222,274,301-2,322

Bab el Mandeb (Strait of)
134–5,188
Bahrain 136,139,244
Ben Bella, Ahmed 155
Benin 163
Benjedid, Chadli 172
Biscay, bay of 52
Black Sea 41–2,47,49,76,190–1
Black September Organization
185,188
El Borma oilfield dispute
153,160
Boumedienne, Houary 160,172
Bourghiba 153
Brazzaville 162
Bulgaria 54,66,113,317
Camp David agreements 175,243
Canary Islands 150
Cape of Good Hope, route of
17,20,22
Carter, Jimmy 140,306
Carter doctrine 300
Carter administration 313,
315
Casablanca group of 1961–62
157,159
Cases:
Anglo-Norwegian Fisheries
Case (1974) 79,97
Anglo-Icelandic Fisheries
Case (1974) 79,97
Corfu Channel Case 82
North Sea Continental
Shelf Cases 834,117,119
Tunisia-Libya Continental
Shelf Case 103
Central African Republic 66
Chad 173
Channel, Arbitration between
France and the U.K. on de-
limitation of the continen-
tal shelf 103
China 159,164

Compagnie Française des
Pétroles (CFP-Total) 3,8,
12–3
Conference on International
Economic Cooperation
(Paris) 166,259
Conference on Security and
Co-operation in Europe 172
Conoco 13
Contiguous zone 81–2,87
Continental Shelf 2,13,82–4,
91–4

Conventions:
Convention and Protocols for
the Protection of the Medi-
terranean Sea against Poll-
ution (Barcelona Convention
1976) 68,114–7,133,249
Action Plan for the Protec-
tion and Development of the
Mediterranean Basin 114
Blue Plan 67
Regional Oil Pollution Com-
bating Centre 65
Protocols 65–6,114–6,133
Continental Shelf Convention
78,82–4,86,93,103–6,117
Conventions on airline high-
jacking 192
Convention on the Law of the
Sea (1982) 42,50,86–104,
126–9,135,167,208
European Convention on the
Suppression of Terrorism
(1976) 193
European Fisheries Convention
(1964) 97
Fishing and Conservation of
the Living Resources of the
High Seas Convention (1958)
77–8,85,105
High Seas Convention (1958)

77-8,84-5,97,105
Regional Convention for Cooperation in the Protection
  of the Marine Environment
  from Pollution among the
  Gulf Arab countries (1978)
  136
Territorial Sea and Contiguous Zone Convention (1958)
  77-8,86,104-5,122,124
International Convention on
  the Prevention of Pollution
  from Ships (The London Convention) 114,116
Montreux Convention (1936) 91

Cuba(n) 281,306
Cyprus 12,42,57-8,65,67,75-6,
  105,133,179,181,188,191,
  196-7,199-200,206,241-2,
  246-7,312,315
  UK Sovereign Bases 42,315
Declaration of Santo Domingo
  (on EEZs in Latin America,
  1972) 97
Dhofari Rebellion 137
Diego Garcia 140
Direct Action (Action Directe)
  183
Djibouti 135,140
Egypt(ian) 6,8,10,12,15-6,20,
  33,36,47,57,66,68,70,75,106
  129-30,133-5,141,158-9,161,
  163,170,179,184,196-8,200,
  209,215,217-8,241,245,248,
  300,307,312,314
  Abu Qir gas field 36
  Aswan Dam 47
  Badr el Din field 12
  Maryout 12
  Nile river and delta 2,12,
  47,57,66,76,151
  Port Said 8,210

Qina 141
Ras Banas 141
Sinai 12,17,20,134,247
Suez, Gulf of 6,10
Suez Canal 17,20,22-3,31,
  41-2,49,64,68,71,196-9
Sumed pipeline 20,22-3
el Temsah gas field 12
al Tina, Gulf and oilfield
  8,12
Western desert 12
ELAS (Greek Leftist insurgent
  group) 196
Elf (SNEA) 3,8
Enclosed or semi-enclosed sea
  101-2
ENI 3
  Agip 3,8,12,13
  Saipem 3
EOKA B 199
Ethiopia 66,135,293,316
  Eritrea(n) 134
Europe(an) 20,26,31,33,41,52,
  61,138,164-5,179-80,182,
  184-6,191,194,207,218-20,
  225-6,239-41,243,246-7,250,
  252-4,256,258-9,273,293-8,
  301,303,306,309,315-6,318,
  320-2
European Community (EC) 53,66,
  70,97,118,165,172,239-61
  European Parliament 239,240
  EC Commission 239,244-5,255
  European Political Coopera-
  tion (EPC) 242,246-8,260
  London Report 248
  European Court of Justice 52
  European Council 243
  Rome Treaty 165,242,254
  The Nine 246-7
  The Ten 246-8
  The Twelve 260
  Venice Declaration (1980)

243

Exclusive Economic Zone (EEZ)
50-2,68,70,97-100,130,208-9,
217

Eniepsa 3

Etap 36

Exxon 3,12-3,31

Falklands 216,222

al Fatah 181,185,189,191,200

Fertile Crescent 151

FLN 181,196

Food and Agriculture Organisation (FAO) 54,65,113

Foreign Military Bases 139-41,
170-2

France (French) 3,10,14,16,27,
42,47,52-3,55,60,62,64,66,
71,75,102-6,114,133,159,165,
169,170,173-4,179,181-2,
184-5,201,209,215-6,220,247,
253,295-7,299,302,320
Corsica 75,179,194
Fos sur Mer 31
Marseilles Oil Terminal 189

Front for the National Liberation of Corsica (FNLC) 189

Gabon 163

Gafsa Raid (1980) 155

Gara Jebilet iron deposits 153

Gas, natural 24-38,41,169

General Council of British
Shipping 189

General Fisheries Council for
the Mediterranean (GFCM) 54,
70,113

Germany (Federal Republic of)
84,188,190,201,215,251,281

Gibraltar 42,70,75,105,122,
168

Gibraltar, Strait of 61,71,75,
91,115,121,169,207,210-11
Cape Spartel 115

Giscard d'Estaing, Valéry 244

Greece (Greek) 6-8,10,13,33,
47,51-2,55,57-8,66-7,70,75,
102,105-7,112,117,127,133,
179,181,189,194,196,206-8,
209,215-6,240,244,246-7,249,
293,297,300-1,312-5,317
Cape Matapan 75
Corfu 107
Crete 60,76,210,211
Evros, Netos and Strimon
rivers 66
Katakolon oilfield 13
Paxoi 13
Rhodes 76
Thessaloniki 7
Zante 107

Greece-Italy agreement for
pollution control in the
Ionian 117

Greece-Italy Delimitation
Agreement (1977) 107,110

Green March 155

Group of 77: 127,166

Gulf, Arabian or Persian 15,
17-9,22-3,26,31,33,37,127,
129-30,135,137-8,140,142,
189-90,207,244,293-300,
302-4,306-9,312-3,315-7,
320-2

Gulf Cooperation Council 131

Gunboat diplomacy 223,275-80

Hassan II 155,184

Hassan-Bel Bella meeting
(1965) 160

Hawar (Island of) 13

High Seas 84-5,94-5,221

Hispanoil 13

Hormuz, Strait of 15,19,137-8,
209

Horn of Africa 138,175-6

Hot Pursuit 85,95

Hussein, Saddam 18,198

Iceland(ic) 97,121

Indian Ocean 37,129,138,140,
  321
Innocent passage 80,82,87-9,
  129-30,138,209
Institut Française des Pétroles 3
Internal waters 78-9,86,131
International Commission for
  the Conservation of Atlantic Tunas 144
International Court of Justice
  8,12,52,79,82-4,97,103,107,
  112,117-20,133,168
  Statute of the Internatio-
  nal Court of Justice 93,
  100
International Energy Agency
  (IEA) 243
International Law of the Sea
  see: Convention on the Law
  of the Sea of 1982
International Maritime Con-
  sultative Organisation 114
International Seabed Authori-
  ty (ISBA) 56,58
Ionian Sea 13,67,75-6,117
IPC pipeline 16
Iran(ian) 17-8,23,31,33,128,
  136-8,142,244,293,298,312,
  314-6,321
  Iranian fields 16
Iran-Iraq Treaty of 1937 on
  boundary delimitation in
  the Shatt-al-Arab 137
Iran-Iraq war 18,243,245,293
Iraq(i) 16-8,22-3,33,129,
  136-8,142,158,196
  Basrah and Kirkuk oilfields
  17-8
Iraqi-Turkish pipeline 22
Irgun 181
Israel(i) 10,12,16-7,20,47,
  75,102,104-5,113,127,130,

133-5,142,179,181-2,184,188,
  194,196-201,207,210,215-7,
  241,243,248,277,300,304,314
Haifa 16,188
Tel Aviv 191
Eilat 188-90
Italy (Italian) 6-8,10,12-4,
  26-7,31,33,36,47,51-3,55,
  57-8,60,62,64,66-7,70-1,
  75-6,104-7,116-7,133-4,
  168-9,179,181,185,215-8,220,
  247,253,295,297,300-2,314,
  317
Aquila field 8
Calabria 107
Gela and Santa Maria "A"
Mare oilfields 7
Lampedusa, Lampione, Linosa,
Pantelleria 76,107
La Spezia 31
Po River 62
Rome 150,184-5,254
Sardinia 75,107,179,212
Sardinian (shelf) 57
Sicily (Sicilian) 2,7,13,
  (53),56,75,107,134,150,
  299,316
Taranto (Gulf of) 107
Trieste (Gulf of) 106,188,
  190,211
Italy-Spain Delimitation Ag-
  reement (1978) 107-8
Italy-Tunisia Delimitation
  Agreement (1971) 107,109
Italy-Yugoslavia agreement
  for pollution control in
  the Adriatic 116-7
Italy-Yugoslavia Delimitation
  Agreement (1968) 106,111
Jamaica 56
Japan(ese) 31,51,138,139,164
Jordan 134,190,196-8,243
Kenya 140

Kuwait 22-3,129,136,244
Latin America(n) 126
Law of the Sea (see Convention on the, 1982)
League of Arab States 127,129, 159,175,243-4,254
Levant 52
  Levantine Basin 76
League of Nations 76
  Charter, 252
  Codification Conference (1930) 77
  International Law Commission 77
Lebanon 12,105,133,141,179, 181,183-4,189-91,194,196-9, 201-2,210,241,243,247,302, 304,316
  Beirut 133,188,199,201,210, 302
  Sidon 16
  Tripoli 16,18,191
Libya(n) 2,6,8,10,12-3,15-6, 31,42,48,52,67,70,75-6,100, 103,106-7,112,119-20,131-3, 142,149-50,154-7,160-2,165-8 170-5,179,181,184-5,189, 200-1,209,214-7,241,245,250, 303,314
  Brega 31
  Libyan Bureau of Foreign Liaison 132
  NC-41 field 8
  Sirte (gulf of) 12,106,131, 132,168,175,201,209
  Tripoli(tanian) 8,149-50, 152,171,201,209
  Cyrenaica 12
Lions, Gulf of (Golfe du Lion) 2,14,75
LNG (liquefied natural gas) 25-7,31,33,38
Lomé Agreement 242,245,255

Luttwak E. 223
Maghreb(i states) 52,156, 159-60,167,169-73,176,207
Makarios, Archbishop 198
Malta 2,10,12-3,42,52,56,58, 65,68,70-1,75,86,97,104-6, 112,133,181,192-3,209,220, 241,247,303
  Maltese Islands 76,97
  Malta Channel 70
  Maltese National Front 189
Marmara Sea 76,91
Masirah (island of) 140
Mauritania 129,133,159-60,200
Medina Bank 12
Mediterranean Action Plan (MAP) see Conventions: Barcelona Convention
Middle East(ern) 16,33,127-30 141,161,179,182,193-4,207, 214,219-20,240,242,246-8, 250,253,293-7,299,302,304-8, 312-4,316-7,322
Mobil 12
Mobutu, Sese Seko 163
Monaco 42,75
Moro, Aldo 183-5
Morocco 10,47,49-51,53,57, 60-1,68,70,75,105-6,114,120, 133,149-50,152,154-63,165, 167-75,179,181,184,191,194, 196-200,214-9,241,245,300, 314
  Casablanca 162
  Fes 150,152
  Point Cires 76
  Rabat 152
  Onarep (Office National des Recherches Pétrolières) 10
Nasser 134
NATO 55,60,207-8,218-20,240, 246-8,253,260,271,273-4,

287,293-8,287,293-8,300-3,
306-7,309-11,314-8,320-1
New York 86
Nigeria 26,162
North Africa(n) 26,66,70,75,
149-151,153-6,159-65,167,
171,173-6,179,185-6,191,194,
214,217,220,240,314
North American Continent 26
North Sea 2,6,97,193
Cormorant field 6
Norton Moore, John 122
Norway 121,193
Jan Mayers Island 121
OAS 181-2
O'Connell, D.P. 76
Oil:
offshore exploration and
production 1-14,41
transportation 15-24,62-4,
67-8
Oman 128,137,140-2,244
Organisation for African Uni-
ty (OAU) 127,159,162-3,174,
197,207
Otranto Channel (Strait) 33,
75,211
Pacific Ocean 58,70
Pahlevi, Shah Mohammed Reza
18,137,312,315
Palestine 151
Palestine Liberation Organi-
sation (PLO) 127,180,184,
201,244
Pardo, Ambassador of Malta 86
Pélerin (drillship) 3
Phalange Libanaise 184
Polisario 50,170,184,189,194,
200,207
Polisario Trail 170
Pollution control 41,61-67,71,
100-1
Popular Front for the Libera-

tion of Palestine (PFLP)
183-5,188,190
Portugal (Portuguese) 50,
241-2,244-6,297,300,312,314
320
Qadhafi, Muammar 131,136,166,
175,184
Qatar 128,136,244
Rapid Deployment Force (RDF)
138,141-2,302-3,309,318
Reagan, Ronald 128,131-2,138
Red Army Faction 182
Red Brigades 183-5
Red Sea 2,10,20,22-3,49,60,
70,127,129-30,134-5,138,
141-2,184,188
Rhone River 62
Romania 54,113,208
Sadat, Anwar 184,200,312
Sahara 155,170,173,179,189,
199,200,245,303
Western Sahara (also Spa-
nish or Moroccan) 153,155,
160,163,199-200,207,248
Sahel 158
SALT II 296
Saudi Arabia 16,19,20,22-3,
128,130,134-6,138,141,190,
244,250,316,321
Dhahran oil fields 16
East-West pipeline 19,22
Jubail 139
Petromin 19
Ras Tanura 19
Yanbu (port/pipeline) 19-20
22-23
Saudi-Sudanese Red Sea Joint
Commission 135
Seabed mining 41,55-61,70-1,
126-8
Senegal 163
Shatt-al-Arab 17-8,23,136-7,
142,304,315

Shell 3,8,10,12
Sherifian Empire 149,155,159
Sicilian Channel 26,37,211-2
South Africa 60,163
South America 97
Somalia 129,134,140-1,
Soviet Union (USSR) (Soviet)
 38,71,134,140,158,170-1,
 173-5,194,206,208,210,215,
 220,225-6,240,245,248,252,
 259,267-89,292,295-303,
 306-8,311-8,320-1
 CPSU Congress (XXV and
 XXVI) 278
 Soviet Armed Forces 276,279
 Soviet Mediterranean Squad-
 ron or Eskadra 269-70,
 272-3,287-9,301,304,318
 Soviet Naval Aviation 219
 Soviet Navy 269,2756
Spain (Spanish) (2),6-8,10,14,
 31,36,42,47,49-53,55,58,60,
 624,66,68,70,75,104-7,114,
 118,133,160,168-9,179,181-2,
 184-5,191,196,199,207,214-6,
 241-2,244-6,261,297,300,312,
 314,317,320
 Baleares 2,7,14
 Balearic Basin 75
 Ceuta 50,60,76,168
 Melilla 50,60,168
 Minorca 107
 offshore oilfields 7
 Point Marroque 76
Steadfastness Front 160,175
Stern Gang 181
Submarines (warships) 12,80,
 82,212,215-6
Sudan 141
Syria(n) 12,16-8,22,66,75,105,
 137,158,179,184,191,196,201,
 210,214-6,243,247-8,307
 Aleppo and Hama 201

Banias (port) 16,18
Tapline (pipeline) 16
Technological progress in
 weapons systems, conse-
 quences of 210-1,222,
 298-300
Territorial sea 79-81,87-9,
 105,127-9,141-2,221
Third World 56,126,141,172-3,
 193,220,224,276-7,279,281-2,
 292,295,301
Tiran (Straits of) 129,130,
 134,209
Transit passage in straits
 89-91,129
Transmediterranean gas pipe-
 line 3,27,33,36,133,169
Treaty on the Prohibition of
 the Emplacement of Nuclear
 Weapons etc. on the Seabed
 123
Treaty of Osimo (1975) 106
Treaty on the Law of the Sea:
 see Conventions
Truman, Harry 313-5
Tunisia(n) 2,6-8,10,13,15-6,
 26,33,36,48,51,53,55,61,68,
 70,75,103,106-7,112,119-20,
 133-4,149-50,154-6,158-62,
 165,167-72,174,179,181,184,
 197,200-1,214-7,241
 Ejele-Skhirra (pipeline) 159
 Gabes (gulf of) 75,112,168
 Kerkennah islands 112
 Miskar gas field 36
 Offshore oilfields 8
 Tell of Tunisia 150
 Tunis 149,150,152,158-9,172
Tumbs 136,137
Turkey (Turkish) 10,22,33,42,
 47,52,57-8,60-1,70,75,102,
 106,112,119,127,133,179,181,
 194,199,206-10,215-7,220,

241,243,246-8,250,254,260,
293,297-8,300-1,303,307,
312-5,317
Turkish Straits 211
Ankara 22,207,240,247
Bosphorus 91
Ceyhan 18
Dardanelles 91,115,307,313
Iskenderun, Gulf of 2,13
Tyrrhenian Sea 2,14,58,71,75
Uganda 66
United Arab Emirates 136,140,
244
United Kingdom 16,42,75,97,
103,104,105,170,193,207,209,
218,240,251,295,296,313
United Nations 76,86,126,189,
197,199,200
Charter 77,87,252
Conference on the Human En-
vironment (Stockholm,
1972) 65
Conference on the Law of
the Sea (UNCLOS):
First (1958) 77
Second (1960) 77,79
Third 55,77,96,100,102-4,
126,129,131,135,249
Environment Program (UNEP)
67,114,116
First Committee of the Gen-
eral Assembly 86
General Assembly 77,86,166,
247
Security Council 132
United States 55,60,71,106,
127-8,131-2,135,138-42,164,
170-1,173,175,192,194,201,
209-10,215,218,220,225-6,240,
243,252,261,273,277-8,280,
282,292-322
Dept. of State 192
Air Force 294

Army Corps of Engineers 139
Cognac field (Louisiana) 3
Dept. of Defence 140
Dept. of State 192
LOOP (harbour, USA) 17
Washington 243,312
5th Fleet 140
6th Fleet 209-10,280,289,
294,299,300-1,316-8,
320-1
Utrecht (Treaty of) 105
Waldheim, Kurt 189
Warsaw Pact 208,219-20,271,
297,303,317
World Bank 36
Yamani, Ahmed Zaki 19
Yemen, North 135,244
Yemen, South 128
Yugoslavia 13,42,53,55,57,67,
75,104,105,106,107,116,181,
185,191,206,215,216,246,254,
300,301
Jabuka, Kajala, Pelagruz
islands 106
Zaire 165
Zimbabwe 60